Japan's International Fisheries Policy

Few nations rely upon the ocean as much as Japan for livelihood, culture and transport. The seas have long played a vital role for the Japanese, helping to support the economic and social life of a nation that possesses few resources and little arable land, and sustain a population that has nearly tripled in the last century. Fish are a distinctive feature of the Japanese diet, constituting nearly half of all animal protein consumed – the highest rate in the world. The industry itself has provided an impetus for coastal community growth and national economic development over the past century, while fisheries have worked their way into Japanese culture and customs, serving as a dominant symbol in traditional arts and folklore.

This book explores the overarching rationale that motivated Japanese international fisheries policy throughout the post-war period until today, highlighting the importance of international fisheries to Japan and the stature this resource has occupied as a national interest. It provides a comparative view of Japanese foreign policy at various ocean conferences, treaty negotiations, bilateral diplomatic initiatives and other maritime relations that constitute ocean policy over half a century, and investigates the domestic constituents of national policy. Roger Smith argues that the rationale for international fisheries policy may be best viewed as deriving from Japan's unique defence strategy for its national interests: comprehensive security. Encompassing non-military elements and most importantly defence of economic interests, *Japan's International Fisheries Policy* provides an interesting case study of how comprehensive security is conceptualized and carried out.

Taking a broad view of Japan's international fisheries policies from 1945 to the present, this book highlights the key trends in policy motives and means throughout the post-war period. As such, it will be of great interest to students and scholars of Japanese studies, international and environmental law, resource management and international relations, as well as to policy makers working in the field.

Roger D. Smith is Associate Professor at the University of Kyushu, Japan, and Adjunct Associate Professor at the University of Western Sydney, Australia.

The Nissan Institute/Routledge Japanese Studies Series

Series Editors:

Roger Goodman, Nissan Professor of Modern Japanese Studies, University of Oxford, Fellow, St Antony's College

J.A.A. Stockwin, formerly Nissan Professor of Modern Japanese Studies and former Director of the Nissan Institute of Japanese Studies, University of Oxford, Emeritus Fellow, St Antony's College

Other titles in the series:

The Myth of Japanese Uniqueness
Peter Dale

The Emperor's Adviser
Saionji Kinmochi and pre-war Japanese politics
Lesley Connors

A History of Japanese Economic Thought
Tessa Morris-Suzuki

The Establishment of the Japanese Constitutional System
Junji Banno, translated by J.A.A. Stockwin

Industrial Relations in Japan
The peripheral workforce
Norma Chalmers

Banking Policy in Japan
American efforts at reform during the Occupation
William M. Tsutsui

Educational Reform in Japan
Leonard Schoppa

How the Japanese Learn to Work
Second edition
Ronald P. Dore and Mari Sako

Japanese Economic Development
Theory and practice
Second edition
Penelope Francks

Japan and Protection
The growth of protectionist sentiment and the Japanese response
Syed Javed Maswood

The Soil
A portrait of rural life in Meiji Japan
Nagatsuka Takashi, translated and with an introduction by Ann Waswo

Biotechnology in Japan
Malcolm Brock

Britain's Educational Reform
A comparison with Japan
Michael Howarth

Language and the Modern State
The reform of written Japanese
Nanette Twine

Industrial Harmony in Modern Japan
The intervention of a tradition
W. Dean Kinzley

Japanese Science Fiction
A view of a changing society
Robert Matthew

The Japanese Numbers Game
The use and understanding of numbers in modern Japan
Thomas Crump

Japan's International Fisheries Policy

Law, diplomacy and politics governing resource security

Roger D. Smith

Routledge
Taylor & Francis Group

LONDON AND NEW YORK

First published 2015
by Routledge
2 Park Square, Milton Park, Abingdon, Oxon OX14 4RN

and by Routledge
711 Third Avenue, New York, NY 10017

First issued in paperback 2017

Routledge is an imprint of the Taylor & Francis Group, an informa business

British Library Cataloguing in Publication Data
A catalogue record for this book is available from the British Library

Library of Congress Cataloging in Publication Data
Smith, Roger, 1970-
  Japan's international fisheries policy / Roger Smith.
    pages cm. – (The Nissan Institute/Routledge Japanese studies series)
  Includes bibliographical references and index.
  1. Fishery management, International–Japan. I. Title.
  SH328.S63 2014
  338.3'7270952–dc23
              2014003832

ISBN 13: 978-1-138-09192-4 (pbk)
ISBN 13: 978-1-138-77523-7 (hbk)

Typeset in Times New Roman
by Taylor & Francis Books

Contents

x   *Contents*

List of illustrations

Preface

A most remarkable statistic in this book is that in Japan fish account for nearly one half of all protein consumed. Those familiar with Japanese restaurants will readily understand this, but such extreme dependence on the products of the high seas has long created serious policy issues for decision makers. For many years guaranteeing stable supplies of food has led to food security being subsumed under the broad category of 'comprehensive security', which also includes national defence in the more conventional military sense, energy security and security of raw material supply.

With Japan being poorly endowed with natural resources and geographically isolated by comparison with most other major powers, few decision makers entertain the idea that the world owes Japan a living. Most have been determined to conduct international economic policy in such a way that Japan should be as invulnerable as possible to threats of interruption to supplies of resources, including most crucially food. At times this has tilted policy in autarchic directions, at other times it has meant integration into global economic arrangements.

So far as supplies of fish are concerned, policies since the end of the Second World War have been constrained to adapt over time to international pressures and to radical changes in the international fisheries regime. In this lucid and meticulously researched book, Roger Smith, who comes from a part of Canada famous for its fisheries, shows how the post-war regime based on *mare liberum* (freedom of the seas) gave way in the 1970s to a far more restrictive, and from the Japanese point of view difficult, state of affairs. Whereas in the earlier post-war period Japanese fishing enterprises experienced problems with neighbouring countries, at least they could expand their operations into the wide oceans. However, with agreements stemming from the United Nations Convention on the Law of the Sea (UNCLOS), Japanese fishing boats faced 200-mile exclusion (or partial exclusion) zones in many parts of the Pacific and elsewhere, where they had been free previously to conduct their operations.

Smith takes the reader through the intricacies of these and subsequent negotiations, in which Japan's overriding priority was to secure its supplies of fish, especially in deep-water fishing areas. Eventually, the self-sufficiency aim was diluted through arrangements in which Japan imported a substantial proportion of its fish requirements.

Having coped as best it could with the radical restructuring of its fisheries policy forced upon it by the UNCLOS negotiations, the Japanese government soon found itself having to deal with a further threat to its perceived interests in the shape of the environmental movement. Even though Japan generally accepted the need to conserve stocks of fish species threatened with serious depletion through over-fishing, the government reacted strongly against environmentalist demands for preservation of certain species.

The series of disputes involved leads Smith into an excellent chapter on whaling. Many critics have been perplexed by the Japanese hard line on whaling, and Japan's continuation of 'scientific' whaling, when the whaling industry now employs a mere few hundred employees and government policy creates immense international opprobrium. While the government justifies its policies in part on the basis of 'preserving tradition', Smith shows that whaling was essentially a post-Second World War phenomenon, and argues cogently that Japan refuses to bow before international pressure on whaling, seeing a potential ban on whaling as a dangerous precedent for further restrictions on its fisheries industry in general.

The Nissan Institute/Routledge Japanese Studies Series began in 1986 and is approaching its 100th volume. The principal aim of the series is to introduce readers to different aspects of Japan (especially but not exclusively in the social sciences), paying particular attention to processes whereby Japan – endowed with the third largest economy in the world – is changing and adapting to newly emerging international conditions.

Arthur Stockwin
Roger Goodman

Acknowledgements

Over the course of writing this book, I have become indebted to many people who helped guide my research and writing, and generously offered their time and effort for interviews and correspondence. First and foremost, I would like sincerely to thank my academic mentor Dr J.A.A. (Arthur) Stockwin, OBE, for his guidance and inspiration. This book has been vastly improved due to his careful reading, patient questioning and insightful suggestions. I would also like to acknowledge the assistance given by Professor Peter Mauch and the University of Western Sydney in providing the necessary momentum to help complete this book project. The Institute for Asian Research at the University of British Columbia and the Sir James Dunn Law Library at Dalhousie University also offered valuable research services and facilities. I have had the good fortune of becoming acquainted with many scholars, professionals, colleagues and government officials – far too many to name here – and would like to thank those who generously shared their time and took the effort to help guide me in the right direction, particularly during the initial phases of my research. Above all, I would like to express my gratitude to Ms Yuki Miyaguchi for showing me the best that Japan has to offer and for renewing my faith that any person can better their inherent circumstances through learning, perseverance and wisdom. All remaining errors are my own.

List of abbreviations

ACJ	Allied Council on Japan
BWU	Blue Whale Unit
CCSBT	Convention for the Conservation of Southern Bluefin Tuna
CITES	Convention on International Trade in Endangered Species of Wild Fauna and Flora
COFI	Committee on Fisheries (FAO)
EEZ	exclusive economic zone
EFP	experimental fishing programme
ESB	Economic Stabilization Board
FAO	Food and Agriculture Organization
FCSA	Food Control Special Account
FCZ	Fishery Conservation Zone
FEC	Far Eastern Commission
GATT	General Agreement on Tariffs and Trade
GDP	gross domestic product
GIFA	Governing International Fishery Agreements
ICCAT	International Commission for the Conservation of Atlantic Tuna
ICJ	International Court of Justice
ICR	Institute of Cetacean Research
ICRW	International Convention for the Regulation of Whaling
IMO	International Maritime Organization
IMS	initial management stocks
INPFC	International North Pacific Fisheries Convention
IOTC	Indian Ocean Tuna Commission
ITLOS	International Tribunal for the Law of the Sea
ITQ	Individual Transferable Quota
IUCN	World Conservation Union
IUU	illegal, unreported and unregulated fisheries
IWC	International Whaling Commission
JFA	Japan Fisheries Association
JFC	Joint Fishery Committee
JICA	Japan International Cooperation Agency

JWA	Japan Whaling Association
LDP	Liberal Democratic Party
MAFF	Ministry of Agriculture, Forestry and Fisheries
METI	Ministry of Economy, Trade and Industry
MITI	Ministry of International Trade and Industry (in 2001 renamed Ministry of Economy, Trade and Industry, or METI)
MOFA	Ministry of Foreign Affairs
MSY	maximum sustainable yield
NAFO	North Atlantic Fisheries Organization
NGO	non-governmental organization
nm	nautical mile (equivalent to 1.15 statute (land) miles)
NMP	New Management Procedure
NPAC	North Pacific Anadromous Commission
NRA	nominal rates of assistance
NRS	Natural Resources Section
ODA	official development assistance
OECD	Organisation of Economic Co-operation and Development
OFCF	Overseas Fisheries Cooperation Foundation
PRC	People's Republic of China
PS	protected stocks
RMP	Revised Management Procedure
RMS	Revised Management System
SBT	southern bluefin tuna
SCAP	Supreme Commander of the Allied Powers (also sometimes known as Supreme Command of the Allied Powers)
SMS	sustained management stocks
SWNCC	State-War-Navy Coordinating Committee
TAC	total allowable catch
TPP	Trans-Pacific Partnership
UN	United Nations
UNCLOS	United Nations Convention on the Law of the Sea (also known previously as the United Nations Conference on the Law of the Sea)
US	United States
USSR	Union of Soviet Socialist Republics
WTO	World Trade Organization

Notes on conventions

Japanese personal names are presented in Japanese form: family name first, followed by given name. The exception to this rule is when a Japanese person is the author of an English-language text, in which case the reference follows Western convention. Macrons are used for long vowels with the exception of commonly understood terms in English, as in Tokyo for Tōkyō.

Introduction

International law, Japanese domestic politics and ocean diplomacy

The year 2010 witnessed a violent escalation in territorial conflicts between Japan and the People's Republic of China (PRC) over the Senkaku (Daoyutai) Islands – a group of uninhibited islets found to the south of Okinawa and to the north of Taiwan. For the first time, an official Japanese coast guard vessel collided with a civilian Chinese fishing liner and the ramming was subsequently leaked to social media outlets, garnering nationwide attention. This clash not only registered the lowest level attained in the historic struggle for control over the islands (over a forty-year period), but it also stoked the volatile passions of nationalism and militarism.

Subsequently, in autumn/winter 2012, Japan and China sent a sizeable group of battleships, cruisers, coast guard cutters and other seafaring vessels to square off in the same waters. Since then, several skirmishes, threats of engagement, radar locking, scrambling of fighter jets and 'open fire' scenarios have broken out between the two nations, which not only tested the resolve of both parties but, perhaps unwittingly, stretched the regional security arrangement in Asia to its limits, including the linchpin US-Japan Joint Security Treaty.

This series of events symbolized the dangerous state of affairs in Sino-Japanese relations and, combined with North Korea's bombing of Yeonpyeong Island earlier in the year, of inter-Asian relations. At stake is not only the issue of the territorial integrity of the island chain, but more importantly, the control of resources in surrounding waters, such as fish, oil, gas and minerals.

The dispute has also led to a revival of interest in the state of maritime law in the region. Practitioners and academics alike have had to review the instruments at their disposal to ensure the non-violent legal resolution of conflict, both territorial and economic, among the parties, while also investigating means to ensure that such disputes do not escalate beyond control in the future.

The Senkaku incidents may very well be a precedent-setting case in the Pacific. Naturally, the level of antagonism and threat of conflict over such a tiny island chain is perplexing and an issue that must be addressed with both seriousness and caution. In previous disputes, the liberal peace theory – whereby economic and domestic political payoffs outweigh the military-legal advantages – have prevailed once the course of the dispute ran its course. This latest encounter was different.

Observers may have been tempted to conclude that legal instruments and diplomatic treaties were somehow insufficient, which led to misunderstandings of territorial definitions, historic cooperative agreements or maritime boundaries. The actual situation is quite the contrary. A comprehensive set of legal instruments, treaties and international law already exists to regulate and contain conflicts in maritime East Asia, while a plethora of administrative bodies help regulate the daily interaction of state actors, such as flagged-fishing vessels at sea and cargo ships navigating ocean lanes. Many avenues and levels of appeal are also provided to mitigate conflict, from the dispute-resolution body contained within the United Nations Convention on the Law of the Sea (UNCLOS) to the International Court of Justice (ICJ).

Instead, political and economic incentives may lie behind the dispute. The 2010–13 clashes demonstrated that the time may have come when patriotism outweighs profit in Sino-Japanese maritime relations. This will not only make the peaceful resolution of disputes even more difficult to achieve in future scenarios of territorial and resource conflict, but it may also suggest that the national political mood may be in favour of brinksmanship and conflict, whether violent, threatening, posturing or defensive. By all accounts, a greater understanding of the interests of the parties involved, and not just the legal context, is merited to ensure that a military clash does not break out in the region. In other words, vital political-economic interests vie for influence, sometimes within and sometimes at the expense of international law and diplomacy.

This suggests that the nature of international maritime conflict in the Pacific is largely the product of *national* interests and other domestic constituencies of foreign policy. Certainly, domestic industries are the primary actors when it comes to fisheries disputes, and it is not a mystery that the fishermen involved in the altercations live near the disputed fishing grounds. Chinese fishermen sailing from nearby Taiwan or Fujian Province are likely to run afoul of Japanese hailing from Okinawa setting nets for the same schools of fish. What is perplexing is how such a localized dispute becomes an international incident that tests the boundaries of international law and security arrangements in the Pacific.

However, the conflict involves more than this. There is also a clash of values pertaining to core norms in foreign policy between Japan and the PRC. Japan has long seen its fisheries and food industries as a matter of national security, elevating the industries to the level of a strategic interest. The PRC, on the other hand, has been more assertive of its ocean claims in recent years since it now has a revamped navy with power-projection capabilities, whereby territorial projection is coupled with testing neighbours' resolve in order to bolster the credibility and image of the national government for its own citizenry without antagonizing important domestic power bases, such as political competitors within the governing politburo.

What is lacking in the resolution of such territorial disputes is not the legal framework for dispute resolution, but the domestic political will to solve outstanding international issues. To put it another way, the political premium derived from conflict can sometimes outweigh the incentive for the conclusive resolution of such disputes. When a dispute escalates into a potential military stand-off, however, then the international community rises to attention. The most important questions of international law and diplomacy relate to how parties might ensure that such conflicts do not spiral out of control into a military confrontation or possible arms race.

Better comprehension of the domestic political and economic incentives behind foreign policy making is one method to ensure better understanding among international antagonists. Various interest groups, industry representatives, political parties, bureaucratic agencies and non-governmental organizations share a stake, of differing proportions, in foreign policy decision making. It is these domestic interests, mediated by international organizations and legal instruments, which help guide the outcome of events in the transnational sphere.

These same national-level political interests, however, served as the agents of codification, enactment and enforcement of international ocean law in the first place. To put it more accurately, those states whose foreign policy has been inspired by more than the sum of its domestic interests and who looked to universal norms, such as cosmopolitanism and conservationism, led the charge in the 1970s for better scientific management of the oceans. The United States, among other nation-states, was – and still is – a powerful proponent that helped set in place the new legal regime, even if it has not yet ratified the UNCLOS treaty domestically.

Ocean governance can be seen as the product of three stages of interaction: 1) international law that sets the context for ocean rights and obligations; 2) national politics, whereby the domestic interest or foreign policy of a nation-state, such as Japan, aims to appropriate, defend and advance its own interests; and 3) international ocean conflict and diplomacy, whereby the state and other actors interact and negotiate under the aegis – guiding principles, written regulations, power balances and ecological limitations – of international law. Each of these will be reviewed in turn, with the weight of consideration in this book in favour of national politics and ocean diplomacy, since this is where the most instructive narratives may be found.

International law

The codification of international law relating to oceans is extensive, expansive and detailed. International waterways, fisheries on the high seas, straddling stocks, piracy, continental shelf and subsoil property rights, territorial waters, riparian and freshwater waterways, transboundary pollution and hazardous waste transport are just some of the areas covered – i.e. defined, described, parametered, jurisdictionalized and authorized – by the various treaties and codes that make up the Law of the Sea. The base documents, UNCLOS, set out initial agreement among parties in 1982 (promulgated in 1994) with regard to legal rights and obligations with respect to the oceans.

This book will not be an overview of the legal provisions as they relate to the Pacific and Asia, since this is more than adequately covered by other academic publications. For a thorough evaluation of contemporary issues in international ocean law, the *Oxford Handbook of International Environmental Law*, edited by Daniel Bodansky, Jutta Brunnée and Ellen Hey, provides a thorough overview of core concepts, analytical tools and future challenges to global environmental and ocean law, while *Maritime Border Diplomacy*, edited by Myon Nordquist and John Norton Moore, details maritime delimitation and boundary disputes with a special focus on Asia.[1] Copies of treaties, laws, acts, case law and resolutions pertaining to ocean law – both national and international – can be found in the *New Directions in the Law of the Sea*, edited by Roy S. Lee and Moritaka Hayashi (including yearly updates).[2]

Instead, this book aims to look at the diplomacy relating to codification, especially as it pertains to Japanese foreign policy and domestic interests, while also highlighting specific legal principles and treaties as they relate to Japanese involvement in various ocean conflicts in the Pacific over a sixty-year timeframe. The most intensive concentration of maritime disputes for Japan came during and after the promulgation of UNCLOS in the 1970s and 1980s, since it was during this time that new boundaries were being defined, fishing grounds claimed, domestic environmental laws promulgated and new petroleum resources discovered.

National politics: the case of Japan

Few nations rely upon the ocean as much as Japan for livelihood, culture and transport. The seas have long played a vital role for the Japanese, helping to support the economic and social life of a nation that possesses few resources and little arable land, and sustain a population that has nearly tripled in the last century.

The Japanese are heavily dependent upon fisheries that constitute an integral part of their diet, economy and culture. Fish are a distinctive feature of the diet, constituting nearly half of all animal protein consumed – the highest rate in the world. The industry itself has provided an impetus for coastal community growth and national economic development over the past century, while fisheries have worked their way into Japanese culture and customs to an extent rarely seen elsewhere, serving as a dominant symbol in traditional arts and folklore. Perhaps it is this reason that many officials refer to Japan as a 'maritime nation' or 'sea people' when passionately defending their nation's right to free access to the sea, or why they became particularly aghast when restrictions were aimed at their nation.[3]

Increasing demand for protein during the last century, due to population growth and diversified consumer desires have rendered the abundant littoral stocks of Japan's coastal waters insufficient to meet the quantity and variety of marine resources required. Rising demand for fish, especially in the post-war period, created an impetus for Japanese fisheries to expand into more distant oceans around the world.

These international fisheries have provided an important source of fish for the domestic market for many decades.[4] Distant-water fisheries grew in importance as they augmented and later superseded domestic supplies of fish, and as fishing ventures expanded overseas, they came into conflict with other nations over control of various fish stocks. The Japanese government consequently had to negotiate international arrangements with regard to rights of access and management of fisheries resources. Fisheries thus became not only an issue of domestic importance but also a matter of foreign policy.

Japan's international fisheries policy has been premised on a strategic motivation to guarantee access to fisheries that were rapidly becoming a vital resource. Attempts to foster such food security largely centred on the goal of promoting self-sufficiency. Thus, international fisheries policy in the post-war period has been domestically motivated by a quest for food self-sufficiency and the success with which Japan was able to achieve this goal was at first facilitated by an international system based upon the principle of freedom of the high seas, but was later constrained by newly developed international norms that granted preferential rights to coastal states for the management of fisheries resources.

This strategic motivation behind an essentially economic issue provides an instructive example of Japan's distinctive security stance in the post-Cold War era – *Comprehensive Security.* As an overarching interpretation of security quite different from the traditional emphasis on strategic and power calculations, comprehensive security places the pursuit of reliable trading partners and routes, economic development and secure access to natural resources on the same level of importance as conventional military notions of security. The case of international fisheries provides an interesting illustration of how the strategy for comprehensive security was conceptualized and carried out in the post-war era. The definition and origins of this term will be examined in greater detail later in Chapter 6. Since the topic of fisheries is conspicuously absent from most discussions on Japanese politics, economics, history and international relations, this study also aims to contribute to a better understanding of the important role that international fisheries have played in Japanese economics and culture.[5]

Ocean conflict and diplomacy in the Pacific

Since the end of the Pacific War in 1946, countless negotiations have been held and concluded with regard to both global treaties and regional arrangements governing fisheries, territories, ocean use and the like. Until the early

1950s, with the passage of Chilean domestic law claiming a 200-nautical mile (nm) territorial sea, freedom of the seas (or *mare liberum* as it is also known) had remained unchallenged as an operating principle for the conduct of global maritime law for centuries. Upon seeing the limitations that a constrictive 3-nm territorial sea imposes upon enforcement under such a system, with no provisions beyond that zone, states initiated a rush in treaty negotiations and domestic legislation of their own to modify global *mare liberum* with their own fishing zones in the 1970s. Such treaties were later used as precedents in subsequent negotiations under the auspices of UNCLOS, so that the Chilean example on fishing zones was later used as a precedent for Japanese diplomats when negotiating treaties with their own neighbours.

This book does not aim to give a overview of the making of the law of the sea nor a history of Japan's negotiating position with respect to this international codification. Overviews of the negotiations leading to the UNCLOS agreement are many, but a particularly insightful analysis can be found in James Harrison's *Making of the Law of the Sea*, while the main source on Japanese negotiations during these conferences is Tsuneo Akaha's *Japan in Global Ocean Politics*.[6] Both are useful starting points for further study.

Instead, this book will provide a comparative view of Japanese foreign policy at various ocean conferences, treaty negotiations, bilateral diplomatic initiatives and other maritime relations that constitute ocean policy over half a century. More importantly, it will investigate further the domestic constituents of national policy, including the incentives, aims and vision that went into the formulation of ocean policy in order to coincide with a greater national agenda. It will be seen in subsequent chapters that two episodes were paradigmatic for the Japanese: the first being the regional fisheries treaties Japan negotiated – with US backing – in the North Pacific in the 1950s and the second being the enclosure movement that effectively placed restrictions on fishing practices from the late 1970s onward.

In the first case, Japan benefited greatly from the application of *mare liberum* precepts in fisheries agreements with the United States, Russia, China and Korea, especially considering the sometimes harsh alternatives that were being considered by its neighbours. Support from the United States at this juncture was essential in setting the groundwork for Japanese post-war fisheries. The enclosure movement reversed some of these gains, by placing restrictions on high seas and overseas fishing operations, and through the effective 'nationalization' of important fishing grounds in the North Pacific and elsewhere. Japan has since been devising a foreign policy aimed at defending its interests in securing fisheries resources through other means, including establishing new joint ventures and espousing food security as a cornerstone of its UN diplomacy.

What does the case of Japan's international fisheries and ocean policy illustrate in terms of the broader question of international law and management of transboundary resources in general?

- International law and norms mediate between domestic demands for food security (i.e. food self-sufficiency) and international constraints placed upon such aims (fears of resource depletion).
- Norms typically arose out of specific contexts, but as they became better defined and understood, they served as precedents in parallel, yet unrelated, cases. An example of this is how the precautionary principle was first devised in West Germany in the 1950s in response to domestic pollution cases, but was later adopted as a key concept during negotiations with Japan for a migratory fish stock agreement in the mid-1970s.
- Fisheries treaties typically set the precedent for other ocean resources with regard to drawing exclusive economic zone (EEZ) boundaries and dispute resolution. Future climate change, however, is most likely to have the most profound effect on the distribution of transboundary fish stocks (i.e. they are most vulnerable to climatic variability and environmental damage), since other ocean resources, such as minerals, oil or shipping are comparatively durable. Thus fisheries will require the greatest flexibility in future management structures.
- Enforcement and adherence: international bodies usually have dispute resolution mechanisms, but little recourse to enforcement of provisions. The responsibility for regulation compliance and enforcement normally falls within national jurisdictions. Codification allows enforcement through legal entities (such as the Magnusson-Packwood Amendment to protect American salmon fisheries), UNCLOS arbitration and the ICJ, which serves as the pinnacle for interstate conflict resolution. Of course, states reserve the right to act unilaterally, including the right to defend their interests through force.
- Treaty architecture: unlike Atlantic and Mediterranean applications of the Law of the Sea, non-formal arrangements predominated in the Pacific.
- The value of fishery resources is rarely reflected in the market price, which suggests that the transboundary nature of the resource complicated management as it pertains to limiting or regulating catches.

Key terms: comprehensive security, food security and self-sufficiency

This monograph is largely concerned with the overarching rationale that motivated Japanese international fisheries policy throughout the post-war period until today. In part, this will be logically deduced from an understanding of the importance of international fisheries to Japan and the stature this resource has occupied as a national interest.[7] This will be done with an eye on a classification test for objectives that may be defined as 'national interest': the actions of leaders must be related to general objectives (i.e. not those of a particular class or group alone) and the ordering of preferences must persist over time.[8] This monograph argues that the motives and strategies of Japanese international fisheries policy pursued national goals with the cooption of sectoral interests, persisted throughout the post-war period despite changing governments, have continually ranked as an important component of Japan's foreign policy, and may thus be seen as an objective of national interest.

Consequently, Japan has adopted a rationale and strategy when devising foreign policy for fisheries. These policies have generally been congruent with US maritime policy, but at key times, when the Japanese government has felt that its interests were at stake over fisheries, it has taken a stance at variance with even their closest ally. International fisheries are an interesting example whereby Tokyo has adopted an independent – if not always successful – strategy on issues that has, at times, differed significantly from Washington, even to the detriment of wider relations. This is best illustrated by the Ministry of Foreign Affairs' (MOFA) and the Fisheries Agency's policy with respect to international whaling and during EEZ negotiations in the 1970s.

The rationale for international fisheries policy may be best viewed as deriving from Japan's rather unique defence strategy for its national interests: comprehensive security. This interpretation of security encompasses non-military elements into its strategic calculations, most importantly defence of its economic interests. Food security, accompanied by fuel security, is considered an important national interest, which is largely measured in terms of self-reliance, but also in terms of secure access to resource supplies, such as open fishing grounds and reliable trading partners. Although comprehensive security is a relatively new term, it defines a concept that has characterized Japanese foreign policy throughout the post-war period.[9]

Comprehensive security highlights the importance of the economic dimensions of security, both how economic instruments may be used to promote stability as well as how a stable economic setting leads to greater security for all. While not ignoring the central position occupied by the Japan-US Security Treaty, comprehensive security nonetheless recognized the economic imperatives for security and supported the role of foreign policy in securing access to resources, safe trading lanes, stable trading partners and fostering future industries as essential elements for a safe and stable world.[10] It is congruent with the style of diplomacy for which the Japanese were conspicuous in the 1950s and 1960s, which sidetracked political impediments in favour of economic considerations and preferences; a style frequently referred to as 'economic', 'value-free' and 'omni-directional' diplomacy.[11] It can be said that the adoption of a comprehensive security strategy in the post-war period changed the way that Japanese policy makers perceived resource scarcity and external threats.

One of the cornerstones of comprehensive security policy has been food security. The term 'security' is used to frame this concept because it is concerned with the stability and continuity of the supply of the nation's foodstuff, especially during times of national emergency or resource shortages. Continuous and stable supplies of staple foods are deemed essential to the health of a society and their importance as a national security issue becomes evident especially when shortages threaten. A key concept underlying food security is a perceived threat of vulnerability to secure sources of supply, such as food crises caused by adverse weather, structural changes in world food markets, political disturbances affecting international trade and food-market disturbances within Japan.

Understanding Japanese foreign policy

Focus will be placed mostly upon the interaction between the Japanese state, largely represented by the bureaucracy, and the international system with its aggregation of international customs and rules. Fisheries policy has been devised within the constraints and freedoms contained in international rights and rules as codified in treaties. Meanwhile, Japanese policy makers had an impact on the development of these international rights and rules since they also served as negotiators at international forums. Thus Japanese international fisheries policy can be seen as an interaction between nationally determined goals and the international system's definitions of rights and rules that shape the capabilities of achieving them.

This monograph will examine Japan's international fisheries policy at the state level, with particular attention paid to the bureaucratic agencies responsible for the development and pursuit of foreign fisheries policy, namely MOFA and the Ministry of Agriculture, Forestry and Fisheries (MAFF) (including its subsidiary organization, the Fisheries Agency). More will be said on the choice for focusing on the central role played by the bureaucracy later this chapter, whereas a closer look at the decision-making constituencies will be offered in Appendix A.

This monograph seeks to examine the motives behind Japan's international fisheries policy and the means used to achieve national goals. This is similar to answering 'why?' and 'how?' questions with respect to comprehensive security policy with respect to food security. In this sense, this monograph is concerned with why Japanese decision makers acted in a particular way and why patterns of action evolved from their decision making. As one study on motivational analysis described this form of inquiry, 'motives are not behaviors but are inferences drawn from behaviors. Hence they are indirectly inferred, *not* directly observed ... motives are postulated as a basis of understanding and are verified by observing behavior'.[12] The behaviours in question being observed are negotiations and policies with respect to fisheries treaties, conventions and international laws. In answering the question of why a particular goal or policy was adopted, the question of how it was pursued can be evaluated in terms of whether the strategy was successful or not in achieving the desired goal.

In order to understand the motives and means of Japan's international fisheries policy, a long-term study of successive negotiations is required to see if patterns develop. The study of a single negotiation case may reveal motivations of actors in a specific time and circumstance, but to ensure that the case is not distinct, a longer-term approach of many cases is required to see if such policy goals and strategies prevail over time and thus may be identified as a consistent motivation of Japanese decision makers. Thus, this monograph will take a broad view of Japan's international fisheries policies from 1945 to the present, with the aim of identifying trends in policy motives and means throughout the post-war period. What the study may lose in terms of depth when compared to a single case study, it will make up in terms of breadth of coverage that can identify policy motivations.

In order to make the examination of more than sixty years of policy and negotiations manageable, this monograph will focus on key points identified within three general periods in post-war history. The periods under consideration are the immediate post-war period[13] during which Japan re-established diplomatic channels with neighbouring countries; the period of ocean enclosure that roughly coincided with the widespread acceptance of EEZs at the third set of law of the sea negotiations within the UN during the mid-1970s to 1980s; and the period of Japanese responses to ocean enclosure that continued from the 1980s to the present. Key points within these periods have been identified by their importance in terms of access to valued fisheries or precedents for international legal norms governing fisheries.[14] Lastly, Japanese policy on whaling will be analysed with an emphasis on current policy in order to assess the policies that are both perplexing and revealing. These key points were testing grounds for Japan's international fisheries policy and set the international legal basis upon which subsequent fishing was permitted and regulated. In this way the monograph can evaluate the policy motivations and means of the Japanese government through an examination of key points during the post-war period.

A wide range of sources were used to examine these key points in depth, which can be divided into historical source materials and current sources. Historical source materials included the US State Department's *Foreign Relations of the United States (FRUS)* and various Supreme Commander of the Allied Powers (SCAP) correspondence, orders, memos and writs that were particularly helpful in piecing together Occupation policy.[15] MOFA, MAFF and national historical archives in addition to various contemporary policy studies, produced by such organizations as the Economic Stabilization Board, helped provide an account of Japanese policy and goals in the immediate post-war and enclosure periods. Bilateral negotiations with the Soviet Union, South Korea and the People's Republic of China in the 1950s and 1960s relied heavily upon *Nihon gaikō shuyō bunsho/nenpyō* (Basic Documents on Japanese Foreign Relations) and Kawakami Kenzō's classic work on Japan's distant water fisheries, *Sengo no kokusai gyogyō seido* (The Post-war International Fisheries Regime).[16] Moreover, the originals of all international treaties, laws and conventions were consulted when the results of fisheries negotiations are under discussion. The treatment of Japan's decision to implement its own fishing zone in 1977 was derived in part from writings of the Director of the Fisheries Division within MOFA, Asomura Kuniaki, who served as a principle adviser and negotiator for distant water fishing policy at that time.

Theoretical context

Apart from the legal works referenced earlier, this monograph addresses two broad themes in current English-language literature on Japan: distant water fisheries and Japanese foreign policy.

Surprisingly little has been written about the Japanese distant water fisheries, given their importance to the economic and social history of Japan. What little literature there is may be divided into two principal categories. The first deals with the subject through the development of international law and Japan's role in legal negotiations, particularly the UNCLOS system that was inaugurated in 1982 and finally ratified in 1994. The most distinguished work in this field is Tsuneo Akaha's *Japan in Global Ocean Politics*.[17] Akaha examines the bureaucratic determination of foreign policy and its outcomes in the context of the 1976 UNCLOS negotiations. This work provides a detailed account of the political manoeuvring of the Fisheries Agency and MOFA in the face of changing international opinion regarding the control of fisheries and the creation of new coastal jurisdictions or the EEZs. While Akaha analyses the shift in legal regimes, he does not examine the impact of the regime shift upon the domestic environment, such as import profiles or the distant water fishing industry. As a consequence of the timing of his publication, Akaha was not able to examine Japan's foreign policy response to this changing regime either.

The second type of literature examines specific fisheries in relation to Japan, such as the tuna and whaling industries. These studies also tend to limit their scope to the domestic realm. The two best known books in this field are Anthony Bergin and Marcus Howard's *Japan's Tuna Fishing Industry: A Setting Sun or New Dawn?* and Arne Kalland and Brian Moeran's *Japanese Whaling: End of an Era?*[18] Both works provide a detailed account of species-specific fisheries, their historical development and, to some extent, the international issues surrounding their management. Neither examines these fisheries in terms of their significance to Japanese foreign policy nor their position within the overarching global fisheries regime. Both of these themes are important for understanding how the system as a whole underwent profound changes at the international level, and the resulting consequences for domestic fishing interests.

Academic research into Japanese foreign policy, on the other hand, is broader and much more comprehensive in its coverage. One type of literature that may be identified are studies that actually look at Japanese diplomacy in terms of intentions, rationale, perceptions and strategies behind why the Japanese do what they do, as distinct from those that focus on processes or policy making.

It is this body of scholarship to which this monograph aims to contribute. Some academics have questioned whether Japan even has a rationale for its policies, whereas others say that foreign policy is designed to 'cope' with problems as they arise with little more than ad hoc reasoning. Most observers agree that Japan must navigate an international system heavily influenced by the United States and thus Tokyo is often obliged to follow Washington's lead.

Scholars such as Gerald Curtis, Michael Blaker and Kent Calder tend to portray Japanese foreign policy as reactive and ad hoc in nature, and generally void of any overarching strategy. In *Japan's Foreign Policy after the Cold War*, Gerald Curtis argues that 'minimalism' is an identifiable feature in Japanese foreign policy. Rather than adopting a proactive policy that sets or substantially influences the international agenda, Japanese foreign policy is largely reactive to trends and crises as they arise: a minimalist attempt to adjust to problems and challenges.[19] Michael Blaker characterizes the Japanese approach to diplomacy as 'coping', by which he means 'carefully assessing the international situation, methodically weighing each alternative, sorting out various options to see what is really serious, waiting for the dust to settle on some contentious issue, piecing together a consensus view about the situation faced, and then performing the minimum adjustments needed to neutralize or overcome criticism and adapt to the existing situation with the fewest risks'.[20] Kent Calder argues in *Crisis and Compensation* that policy innovation in Japan is principally the reaction to crisis, and not interest-group lobbying or the strategic planning of the state.[21] In a similar vein, J.A.A. Stockwin argues that whereas Japan is famous for its industrial and commercial dynamism, Japanese political decision making can often appear 'immobilist' and resistant to change and external pressures.[22]

A variation of this theme is developed by Takashi Inoguchi and Purnendra Jain. They suggest that although Japan is reactive to a restraining international system, it still exercises latitude in foreign policy decision making and strategy, even if somewhat limited. Adopting a mode of argument employed in an earlier book with respect to domestic politics, they suggest that the process of devising and pursuing foreign policy in Japan is analogous to karaoke, in what they term 'karaoke diplomacy'.[23] MOFA and other agencies, in conjunction with politicians, have latitude for choice, much as a karaoke singer may choose from among a range of songs on a karaoke machine. Moreover, the way in which the policy is presented or acted upon may vary in as many ways as one's singing performance. However, the options are always contained within parameters generally sanctioned by Japan's chief allies, most importantly the United States, much in the same way that a singer cannot very well choose a song not contained in the karaoke songbook.

In short, this type of literature argues that since Japan lacks an international strategy, it does not occupy a political role on the world stage commensurate with its economic strength. Instead, Japan's chief ally, the United States, largely shapes the international environment both purposively and through design, and Japan follows its lead or adapts according to each circumstance. Thus Japan has been caught in predicaments, such as in the case of negotiations over textile trade liberalization in the General Agreement on Tariffs and Trade (GATT) or UNCLOS conferences concerned with rule making for the oceans, whereby MOFA and other national representatives were ineffective in defending Japan's national interests, mostly due to lack of foresight, planning and negotiating competence.

A more extreme variation of this theme argues that Japan lacks any distinguishable foreign policy whatsoever. As Jean-Pierre Lehmann once argued, 'there is no public opinion in Japan on foreign policy and ... there is no foreign policy'.[24] Instead, he suggests that Japanese policy pronouncements are sutra-like, repetitive banalities about peace and prosperity. Karel van Wolferen is another scholar who argues that Japanese policy lacks any coherence. He compares the Japanese policy-making system to a 'truncated pyramid' that has no single leading institution responsible for national policy or decision making in emergencies. Instead, political power is dissolved among various semi-autonomous, self-aggrandizing groups, resulting in the degeneration of a line of command.[25]

A very different theme in the study of Japanese foreign policy argues that not only does Japan pursue a rational foreign policy, but some degree of international leadership as well. In *Japan's Foreign Policy for the 21st Century: From Economic Superpower to What Power?* Reinhard Drifte describes Japan as a 'comprehensive power' and economic great power in Asia, and suggests that Tokyo exerts leadership by 'stealth', or in other words, leadership in an incremental and low-profile manner.[26] This monograph is similarly developed by Glenn Hook, Julie Gibson, Christopher Hughes, and Hugo Dobson in *Japan's International Relations: Politics, Economics and Security.* Here, Japan's foreign policy is characterized as 'quiet diplomacy', employing 'a range of consistently low-risk and low-profile international initiatives' to achieve the goals set out by a 'normal state'.[27] This book tends to agree with these findings and suggests that while its international fisheries policies may not exert a leadership function, Japan has certainly influenced the international system concerned with governing maritime resources, and has been effective in achieving most of its foreign policy goals in a low-profile manner as suggested above.

Organization of the book

Chapter 1 will examine the origins and development of Japan's food security policy from the Meiji era to the Pacific War within the context of debates relating to international legal norms and freedom of the seas. The rationale with which Japan's ocean fisheries developed in the post-war period, including the negotiation and adoption of international treaties, was largely influenced by the food security planning system that predated it. *Mare liberum* offered few limitations to inhibit the expansion of fishing effort and can thus be said to have aided Japan in its efforts to become self-sufficient in food supplies.

Chapter 2 will then look at the situation of Japan during the Occupation as it began to normalize relations after the Pacific War. Japan was quite prudent in navigating its way around potentially hazardous regime alternatives that might have been far more restrictive on their activities. Questions of post-war planning and food security were also contemplated at this time and their recommendations were to have a lasting impact upon post-war planning and development trajectories.

Chapter 3 will examine how food security planning mandated adherence to pre-war concepts of *mare liberum* despite greater calls for a restrictive maritime zone. Japan, the United States and Canada concluded Japan's first post-war fisheries treaty, the International North Pacific Fisheries Convention (INPFC), and it purposefully reaffirmed the principle of freedom of the seas upon which other subsequent bilateral fisheries agreements in the North Pacific were negotiated. In effect, it helped reaffirm the open oceans system in the post-Occupation period that facilitated the rehabilitation and eventual expansion of Japanese distant water fisheries. Subsequent negotiations with the Soviet Union, the PRC and South Korea were beset by many problems, such as the absence of formal governmental ties, wartime resentment or the absence of a peace treaty, but each of these barriers was surmounted in turn. Frequently, Japanese industry representatives acted as intermediaries and were able to offer various economic cooperation agreements in conjunction with much coveted fisheries treaties.

Chapter 4 will explain why and how the international regime for fisheries changed its focus away from *mare liberum* in favour of coastal-state stewardship of marine resources. Distant-water operations from the 1970s onward faced an increasingly restrictive environment – known as the enclosure movement – which placed limits, and in some cases total moratoriums, on Japanese fishing activities in international waters. Hereafter, coastal states obtained a greater say in fisheries resource management decisions internationally. Japanese maritime interests felt besieged since their goal of food security was substantially undermined with the reversal of ocean regimes toward restrictions and moratoriums on their fishing operations.

The rising trend toward environmental protection and its impacts on Japan's international fisheries will be examined in Chapter 5. The enclosure movement was not the only international trend that arose to challenge the notion of freedom of the seas. The 'precautionary principle' – a legalistic notion that suggested that environmental risk should be minimized, akin to the expression 'better safe than sorry' – began to gain currency in the 1960s and impel diplomatic negotiations in favour of resource protection and fisheries restrictions. Concomitantly, several developments occurred that would have the net effect of restricting Japanese fishing activities in international waters. The most important of these movements included efforts to ban whaling and driftnets, attempts to have bluefin tuna stocks protected as an endangered species, adoption of a moratorium on salmon catches in the Pacific high seas, pressure to stop fishing in the Bering Sea 'Doughnut Hole', and extending protection of straddling stocks and highly migratory fish in international jurisdictions. These moratoriums, precautionary approaches and extensions of coastal state controls will be investigated in turn to see how they further undermined international support for freedom of the seas and Japan's open access to distant water fisheries.

Chapter 6 examines how a new normative legal movement with regard to the precautionary movement in conjunction with an international society more sympathetic to the aims of environmental protection led to a regime shift in ocean politics. Facing an increasingly restrictive treaty network, Japanese policy makers were once again confronted with the problem of how to guarantee a stable supply of fisheries for Japanese consumption in the absence of the influential backing of the United States. Alone in its pursuit of a new fisheries strategy in the post-EEZ era, Japan re-evaluated the motives behind the earlier fisheries strategy in light of a changed domestic and international environment. A new *comprehensive security* strategy was adopted, which reaffirmed food security and self-sufficiency as one of the cornerstones of the new policy. Although the motives for international fishing policy remained focused on self-sufficiency, new means were adopted to realize this goal.

Next, Chapter 7 will examine how Japan adopted a new fisheries strategy to cope with the increasingly restrictive international environment with respect to fisheries. This new strategy can be divided into four main parts that involved developing coastal fisheries more intensively, increasing imports, negotiating new bilateral agreements and promoting open access to fisheries at multilateral forums. In general, Japan seeks counter-measures that limit fishing effort and restrict fishing grounds while promoting the cause of expanded access to sustainable fisheries worldwide in an effort to create a fairly balanced degree of food self-sufficiency.

The rather unique case of Japanese whaling in ocean law and politics will be investigated in Chapter 8. A review of Japanese whaling policy for the past several decades tries to offer a distinctly political explanation for Japan's sometimes perplexing pro-whaling stance at the International Whaling Commission (IWC). As a single, compact issue that is widely familiar, Japan's whaling policy merits special consideration since it encapsulates the major themes of Japan's larger international fisheries and food policy. It also highlights the contemporary debate regarding the scientific management and sustainable use of natural resources.

The book will offer an analysis of current trends in Japanese food security and comprehensive security policy in Chapter 9. An overview of the goals of current agricultural and fisheries policy will also evaluate congruence with the original foundations upon which the food system has been developed. Lastly, several contemporary issues, such as the Fukushima disaster's impact on the food system, will be covered here.

Finally, the Conclusion will offer a summary of the main thesis with regard to international legal norms and ocean diplomacy, Japanese resource management and food security, and lessons in policy making. Moreover, several new issues and controversies will be highlighted as a means to suggest potential themes of extended research. As can be seen in this case of fisheries, resource diplomacy is an intricate balancing act that will become even more difficult as climate change ushers in stormy waters of even greater proportions. Appendix A offers an outline of the distant-water fishing industry as well as the major players in domestic policy making.

Notes

1 Daniel Bodansky, Jutta Brunnée and Ellen Hey, eds, *Oxford Handbook of International Environmental Law*. Oxford: Oxford University Press, 2007; and Myon Nordquist and John Norton Moore, eds, *Maritime Border Diplomacy*. Centre for Oceans Law and Policy. Leiden: Martin Nijhoff Publishers, 2012.
2 Roy S. Lee and Moritaka Hayashi, eds, *New Directions in the Law of the Sea*. New York: Thomson Reuters/West, 2012, with updates.
3 See, for example, Kate Barclay's 'Ocean, Empire and Nation: Japanese Fisheries Politics,' in *Water, Sovereignty and Borders in Asia and Oceania*, Devleena Ghosh, Heather Goodall and Stephanie Hemelryk Donald, eds. London: Routledge, 2009, 38. Professor Barclay offers a convincing argument with respect to how a pervasive identity of a 'sea people' (*kaijin*) strongly informs Japanese officials and influences national policy with respect to overseas relations and ocean policy.
4 International fisheries are defined as both the distant-water expeditions that have plied the world's oceans as well as imports from fishing operations in foreign waters.
5 With very few exceptions, Japanese fisheries and their international operations have received surprisingly little attention in academic literature despite the fishing industry's importance to national politics and economics. Even such seminal works as John Dower's *Embracing Defeat* and Kent Calder's *Crisis and Compensation* regrettably make no mention of fisheries despite these authors' remarkably thorough treatment of other issues in the post-war period. John Dower, *Embracing Defeat: Japan in the Wake of World War Two*. New York: W. W. Norton, 2000; and Kent Calder, *Crisis and Compensation: Public Policy and Political Stability in Japan, 1949–1986*. Princeton, NJ: Princeton University Press, 1988.
6 James Harrison, *Making of the Law of the Sea: A Study in the Development of International Law*. Cambridge Studies in International and Comparative Law. Cambridge: Cambridge University Press, 2011; and Tsuneo Akaha, *Japan in Global Ocean Politics*. Honolulu: University of Hawaii Press and Law of the Sea Institute, 1985.
7 While fisheries resources may not contain the same strategic value that oil or energy may have, they are nevertheless an essential food supply that if interrupted or cut off for political, military or economic reasons, could lead to dire ramifications on Japanese livelihoods, as was raised as a concern after Japan lost many traditional fishing grounds after the introduction of the 1976 EEZ system. In short, unstable supplies and prices of important resources can upset the general functioning of the economy and strain the political system, and thus Japan has prioritized efforts to secure and stabilize both.
8 This study is particularly instructive on the theme of national interest and strategic resources. Stephen Krasner, *Defending the National Interest: Raw Materials*

Investments and U.S. Foreign Policy. Princeton, NJ: Princeton University Press, 1978. This study is particularly instructive on the theme of national interest and strategic resources. Krasner argues that some resources, such as oil and copper in the case of the United States, are of such importance to the well-being of society that they become a fundamental motive for foreign policy, i.e. a matter of 'national interest'. Although the loss of supply or extreme price variations for such resources may not directly threaten the territorial or political integrity of a state, as might a conventional military threat, foreign policy questions concerned with economics involve more complicated calculations about various aims and are tangentially related to protecting a nation's national interest (see p.13).

9 The term 'comprehensive security' was first coined in 1978 by the Nomura Research Institute, although it initially derived from a proposal by the Japan Socialist Party for the establishment of a new national security council. It gained its greatest currency after the publication of the Inoki Task Force report in July 1980, entitled *The Report on Comprehensive National Security*, which was commissioned by former Prime Minister Masayoshi Ōhira and subsequently endorsed by Prime Minister Zenkō Suzuki (John W.M. Chapman, R. Drifte and I.T.M. Gow, *Japan's Quest for Comprehensive Security*. London: Frances Pinter, 1983, xi–xviii).

10 Another feature of comprehensive security worth noting is that the term was devised at a time when the United States was pressuring Japan to boost its security contributions. Including economic instruments such as overseas development assistance into their security calculations enabled Japan to claim larger expenditures on defence without a sizeable increase in the national budget for military base or weapon procurement.

11 Chapman *et al.*, *Japan's Quest for Comprehensive Security*, 92. Japan started to view potential threats differently in the 1970s in light of its experience with the way non-military crises could cause greater damage to its interests than the military threat posed by the neighbouring USSR. Thus, the term comprehensive security was coined at a time when resource scarcity seemed to pose a greater threat to the well-being of Japanese society than did military insecurity. The two oil shocks, the soybean crisis and the loss of important fishing grounds in the 1970s confronted Japan with the kind of havoc that could be created through resource instability and scarcity, whereas the Vietnam War was a valuable lesson in the limits of a military strategy to respond to external threats. How to secure access and control to vital resources and other economic inputs was hereafter explicitly discussed in Japan. Until this time, comprehensive security was a largely understood but often unmentioned aspect of Japanese foreign policy.

12 Richard C. Snyder, H.W. Bruck and Burton Sapin, 'Motivational Analysis of Foreign Policy Decision-making', in *International Politics and Foreign Policy: A Reader in Research and Theory*, James N. Rosenau, ed. New York: Free Press of Glencoe, 1961, 248, emphasis in original.

13 Including the Occupation (1945–52), during which time Japan was not autonomous.

14 Sources that proved invaluable to identify key points for international fisheries negotiations included the Fisheries Agency's *Wagakuni no suisan gaikō ni tsuite* (Fisheries Diplomacy of Japan). Tokyo: Suisanchō, 2003; and *Suisanchō 50 nenshi* (Fifty-Year History of the Fisheries Agency). Tokyo: Suisanchō50 nenshi kyōkai, 1998; along with Kawakami Kenzō's *Sengō no Kokusai Gyogyō Seido* (The Post-war International Fisheries Regime). Tokyo: Dai Nihon Suisankai, 1972.

15 US Department of State, *Foreign Relations of the United States*. Washington, DC: US Department of State, miscellaneous years; and various SCAP publications and documents such as *GHQ/SCAP Top Secret Records*. Tokyo: Haku Shobō, 1993; and Edward Ackerman, *Japan's Natural Resources and Their Relation to Japan's Economic Future*. Chicago: Chicago University Press, 1953. The Natural Resources Section was charged with fisheries policy during the Occupation.

18   *Introduction*

16 *Nihon Gaikō Shuyō Bunsho/Nenpyō* (Basic Documents on Japanese Foreign Relations, Vols 1 and 2), Kajima Heiwa Kenkyūjo, ed. Tokyo: Hara Shobō, 1984; and Kawakami, *Sengo no Kokusai Gyogyō Seido.*
17 Akaha, *Japan in Global Ocean Politics.*
18 Anthony Bergin and Marcus Howard, *Japan's Tuna Fishing Industry: A Setting Sun or New Dawn?* New York: Nova Science, 1996; and Arne Kalland and Brian Moeran, *Japanese Whaling: End of an Era?* London: Curzon Press, 1992.
19 Gerald Curtis, 'Introduction', in *Japan's Foreign Policy after the Cold War: Coping with Change*, Gerald Curtis, ed. London: M.E. Sharpe, 1993, xv.
20 Michael Blaker, 'Evaluating Japanese Diplomatic Performance', in *Japan's Foreign Policy After the Cold War: Coping with Change*, Gerald Curtis, ed. New York: M.E. Sharpe, 1993, 3.
21 Calder, *Crisis and Compensation*, 20.
22 J.A.A. Stockwin, Alan Rix, Aurelia George, James Horne and Daiichi Ito, *Dynamic and Immobilist Politics in Japan.* London: Macmillan Press, 1988, 325.
23 Takashi Inoguchi and Purnendra Jain, *Japan's Foreign Policy Today: A Reader.* London: Palgrave-Macmillan, 2001.
24 Jean-Pierre Lehmann, 'Japanese Attitudes toward Foreign Policy', in *The Process of Japanese Foreign Policy: Focus on Asia*, Richard L. Grant, ed. London: Royal Institute of International Affairs, 1997, 136.
25 Karel van Wolferen, *The Enigma of Japanese Power: People and Politics in a Stateless Nation.* London: Macmillan, 1989, 41.
26 Reinhard Drifte, *Japan's Foreign Policy for the 21st Century: From Economic Superpower to What Power?* London: Palgrave Macmillan, 1998, 172–73.
27 Glenn Hook, Julie Gibson, Christopher Hughes, and Hugo Dobson, *Japan's International Relations: Politics, Economics and Security.* London: Routledge, 2001, 71.

Bibliography

Ackerman, Edward. *Japan's Natural Resources and their Relation to Japan's Economic Future.* Chicago: Chicago University Press, 1953.
Akaha, Tsuneo. *Japan in Global Ocean Politics.* Honolulu: University of Hawaii Press and Law of the Sea Institute, 1985.
Barclay, Kate. 'Ocean, Empire and Nation: Japanese Fisheries Politics', in *Water, Sovereignty and Borders in Asia and Oceania*, Devleena Ghosh, Heather Goodall and Stephanie Hemelryk Donald, eds. London: Routledge, 2009.
Bergin, Anthony and Marcus Howard. *Japan's Tuna Fishing Industry: A Setting Sun or New Dawn?* New York: Nova Science, 1996.
Blaker, Michael. 'Evaluating Japanese Diplomatic Performance', in *Japan's Foreign Policy After the Cold War: Coping with Change*, Gerald Curtis, ed. New York: M.E. Sharpe, 1993.
Bodansky, Daniel, Jutta Brunnée and Ellen Hey, eds. *Oxford Handbook of International Environmental Law.* Oxford: Oxford University Press, 2007.
Chapman, John, R. Drifte and I.T.M. Gow. *Japan's Quest for Comprehensive Security.* London: Frances Pinter, 1983.
Curtis, Gerald, ed. *Japan's Foreign Policy after the Cold War: Coping with Change.* London: M.E. Sharpe, 1993.
Drifte, Reinhard. *Japan's Foreign Policy for the 21st Century: From Economic Superpower to What Power?* London: Palgrave Macmillan, 1998.

Fisheries Agency. *Suisanchō 50 nenshi* (Fifty-Year History of the Fisheries Agency). Tokyo: Suisanchō 50-nenshi kyōkai, 1998.

——*Wagakuni no suisan gaikō ni tsuite* (Fisheries Diplomacy of Japan). Tokyo: Suisanchō, 2003.

Harrison, James. *Making of the Law of the Sea: A Study in the Development of International Law*, Cambridge Studies in International and Comparative Law. Cambridge: Cambridge University Press, 2011.

Hook, Glenn, Julie Gibson, Christopher Hughes and Hugo Dobson. *Japan's International Relations: Politics, Economics and Security*. London: Routledge, 2001.

Inoguchi, Takashi and Purnendra Jain. *Japan's Foreign Policy Today: A Reader*. London: Palgrave-Macmillan, 2001.

Kalland, Arne and Brian Moeran. *Japanese Whaling: End of an Era?* London: Curzon Press, 1992.

Kenzō Kawakami. *Sengō no Kokusai Gyogyō Seido* (The Post-war International Fisheries Regime). Tokyo: Dai Nihon Suisankai, 1972.

Krasner, Stephen. *Defending the National Interest: Raw Materials Investments and U.S. Foreign Policy*. Princeton, NJ: Princeton University Press, 1978.

Lee, Roy S. and Moritaka Hayashi, eds. *New Directions in the Law of the Sea*. New York: Thomson Reuters/West, 2012.

Lehmann, Jean-Pierre. 'Japanese Attitudes Toward Foreign Policy', in *The Process of Japanese Foreign Policy: Focus on Asia*, Richard L. Grant, ed. London: Royal Institute of International Affairs, 1997.

Nordquist, Myon and John Norton Moore, eds. *Maritime Border Diplomacy*. Centre for Oceans Law and Policy. Leiden: Martin Nijhoff Publishers, 2012.

Snyder, Richard, H.W. Bruck and Burton Sapin. 'Motivational Analysis of Foreign Policy Decision-making', in *International Politics and Foreign Policy: A Reader in Research and Theory*. James N. Rosenau, ed. New York: Free Press of Glencoe, 1961.

Stockwin, J.A.A., Alan Rix, Aurelia George, James Horne and Daiichi Ito. *Dynamic and Immobilist Politics in Japan*. London: Macmillan Press, 1988.

van Wolferen, Karel. *The Enigma of Japanese Power: People and Politics in a Stateless Nation*. London: Macmillan, 1989.

1 *Mare liberum* and the pre-war origins of food security in Japan

This chapter will examine the origins and development of Japan's food security policy from the Meiji era to the Pacific War (1867–1940) within the international legal context of freedom of the seas (*mare liberum*). A recurring theme in administrative and political thinking throughout this period was a belief in the necessity of fostering national self-sufficiency in the domestic production of food. Although not unique to Japan, self-sufficiency thinking did create an incentives system that helped determine the sometimes complex patterns of food distribution, such as rice and fisheries production, within a global oceans regime that imposed few strictures. Early debates within governing circles and their outcomes in terms of policies will be examined to analyse the reasons why self-sufficiency was promoted as a national policy and the means by which this was done. It will be seen that food security planning had an impact upon the development of Japan's ocean fisheries in the post-war period as well as the later negotiation and adoption of international fishing treaties. *Mare liberum* offered few limitations to inhibit the expansion of fishing effort and can thus be said to have aided Japan in its efforts to become self-sufficient in food supplies.

Pre-war ocean regime of freedom of the seas (*mare liberum*)

A longstanding global oceans tradition of 'freedom of the seas' was already firmly established by the time Japan's fisheries extended operations into the high seas in the early 1900s. International maritime and fisheries law was predicated upon the principle of *mare liberum* as originally outlined in Hugo Grotius's 1609 publication of the same name. This principle maintained that because the seas were inexhaustible and inappropriable, and consequently did not share the qualities of private property, no state or navy could own them. Therefore, the high seas were open to everyone's use.[1]

Over the years, the *mare liberum* concept was modified with the widespread acceptance of 'territorial seas'. As a form of *mare clausum*, the territorial sea principle allowed littoral states to claim a narrow 3-mile zone of coastal water ownership.[2] Outside this zone, however, freedom of the seas prevailed.[3] This global regime had maintained consensus for over three centuries.

Consequently, no nation could claim exclusive rights to fish stocks in their natural habitat. Everyone was entitled to harvest fish on the high seas, except within the narrow confines of territorial seas. In economic terms, fisheries resources were allocated upon the basis of an open-to-entry common property system.[4] This was coupled with a procedural device known as the Law of Capture, which provided that the capture of any fish from the ocean automatically transformed the common resource into the private property of the fisherman.[5] This ensured that an economic incentive existed for fishermen who invested in expensive high-powered vessels and diesel engines in order to extend their operations into fishing grounds on the high seas.

Food security and self-sufficiency

Early Meiji-era planners considered several alternatives with regard to economic planning in the service of national development. One of the primary concerns centred on securing food supplies for a growing populations in a country not bestowed with plentiful natural resources, which is otherwise known as the *shigen mondai* in Japanese.

The term 'security' is used to frame the concept of food resources because government planners became concerned with the stability and continuity of the supply of the nation's foodstuffs, especially during times of national emergency or resource shortages. Continuous and stable supplies of staple foods were deemed essential to the health of a society and their importance as a national security issue becomes evident especially when shortages threaten.

A key concept underlying food security is a perceived threat of vulnerability to secure sources of supply that might disrupt international trade and food provision within Japan. Elevating this concern to the state level as a 'national interest' creates a strong rationale for governmental intervention, sponsorship and control of food resources.

Food security can be achieved when the national demand for food is adequately supplied by domestic production, international imports or, more commonly, a combination of the two. Domestic production is commonly measured in terms of self-sufficiency, or the amount of domestic national demand that is supplied by domestic sources as opposed to foreign ones. Self-sufficiency can be determined by various measurements: it can be defined as the *value* of domestic production of food as a percentage of total supplies, or, more commonly, as the relation between the production of original *food energy* in calorific terms (crop production, roughage and domestic fish landings converted into the caloric equivalent of tons of grain) and domestic production plus all imports of crop products and fish (also converted into calorific terms), and the grain equivalent required to produce livestock products that are imported.[6] Self-sufficiency can also be measured in terms of the *weight* of domestic production relative to imported foodstuffs, although this is less commonly used in official statistics.

Self-sufficiency is often expressed as a ratio or percentage, and the lower this figure is, the less 'secure' a nation's food self-sufficiency is deemed to be. Self-sufficiency is deemed desirable because having a high level of self-reliance supposedly means that a nation can feed itself even during times of crisis, without having to rely upon foreign supplies that can be unreliable in the event of excessive international demand or worldwide shortages, or susceptible to be used as a trade weapon in the event of hostilities.

The downside, however, is that this typically incurs substantial costs borne by consumers due to lack of choice, higher prices and markets that are less flexible to individual demand. The incentives system that centralized control of food production – an admittedly important concern for almost everyone – creates an incentive system that can lead to incontestable (i.e. systematic) corruption and abuse.

Food security, however, can also be achieved when stable and reliable supply chains are established with the international market. For this reason, some analysts argue that the promotion of free trade ensures a stable food supply more effectively than policies frequently endorsed by those who seek greater self-sufficiency rates and derive benefits from a protected industry. Only when a national economy becomes dependant on too few sources of supply, it is argued, might vulnerability pose a problem for food supplies, in which case diversification of overseas sources is recommended. Proponents argue that free trade ensures that food prices are generally lower due to expanded competition, products are diversified according to tastes, and supply linkages can be expanded beyond a limited domestic base, all of which create a more stable situation for food consumption. Potential costs exist, however, in terms of fostering local employment and businesses in the food industries.

A longstanding debate between the proponents of both philosophies has existed within Japanese policy-making circles throughout Japan's modern history, which peaked in the early Meiji era and again in national security policy debates in the late 1970s. These debates have followed a similar pattern, whereby the policies required for one course often ran counter to the goals of the other. To increase national self-sufficiency, Tokyo frequently offered a combination of governmental subsidies, tariff protection and investment incentives to promote domestic industries even when these industries were inefficient or unprofitable. Policies that rely upon the economic benefits of free trade, however, typically require the reduction of these very same subsidies and tariffs that create barriers to trade. This was especially true in the post-war period when international trade agreements, such as the General Agreement on Tariffs and Trade (GATT) and the World Trade Organization (WTO), formalized the rules requiring nations to reduce such barriers as a precondition for membership to these organizations.

Policy makers in Tokyo have consistently favoured the policy of self-sufficiency in food production over and above free trade as a national food security strategy, both in pre- and post-war Japan. This is especially true in the case of rice. The Ministry of Agriculture (later known as the Ministry of Agriculture, Forestry and Fisheries, MAFF) has continually adopted policies of subsidies, price supports, trade protection and tax incentives in an effort to boost and maintain a high level of self-sufficient rice production throughout the pre-war period. Indeed, rice policy in Japan would later guide policy making for other food industries as well.

The same goal of self-sufficiency was also applied to other sectors under the Ministry's control, including fisheries production. Self-sufficiency planning permeated official thinking within the Fisheries Agency and served as the conceptual basis for a policy of fisheries expansion in the early post-war era. As will be seen in Chapter 3, this desire for a domestically self-sufficient fishing industry endorsed the reasoning behind Japan's support for freedom of the high seas in international negotiations with neighbouring countries in the early post-war years.

This chapter will first examine the origins of self-sufficiency planning in Japanese food policy from the Meiji era to the post-war period in an effort to identify continuities in policies and official thinking at critical junctures in Japanese history. Self-sufficiency appeared as an early goal of government planners and support for this idea as a national interest has continued until the present day. Indeed, food security policy has led to the development of sectoral interests that, in turn, perpetuated national policy trajectories in a relationship that may be best described as *path dependent*. When possible, connections between food security planning and the development of the distant water fishing industry will also be examined.

The origins of self-sufficiency in Japanese food policy

Japanese policy makers have long grappled with the question of how to secure food and material resources for an expanding economy engaged in an intensive modernization effort. As a country with poor endowments of natural resources and arable land, Japan has struggled with several alternative policy options to overcome what is traditionally referred to as the 'resource problem' (*shigen mondai*). An important component of the resource problem was finding a reliable source of food for a population that was booming and becoming increasingly urbanized; a policy that would later be coined as food security.

Food security – in particular, fostering self-sufficiency in food production – first emerged as an issue of national policy in the latter half of the Meiji era. Although the first Meiji-era leaders placed much faith in markets to supply the nation's appetites sufficiently, subsequent elites started to envision a much greater interventionist role of the state in providing for the nation's needs. Indeed, whereas free trade predominated in policy making in the early Meiji era and had many influential proponents such as Fukuzawa Yukichi, protectionism later gained a hold as nationalists won political gains with the promulgation of the Imperial Constitution in 1889 and victory over China in 1895.[7] It was during this time that 'national policies', along with concomitant economic planning, replaced *laissez-faire* thinking and became a hallmark of government efforts to modernize, principally along the lines of *fukoku kyōhei* ('wealthy nation, strong army'). Protectionist philosophy, as argued by such contemporary intellectuals as Wakayama Norikazu, Sugi Kōji and Ushiba Takuzō, emphasized the importance of fostering domestic industries, even at the expense of the potential benefits of free trade in the short run, in order to strengthen the national economy as well as foster self-reliance in resources key to development and industrialization, including food production.[8] Mobilization of resources for the 'national' effort is the key conceptual argument of such protectionists.

Of particular concern to government leaders in the 1890s was a trend whereby Japan was importing an increasing amount of rice whereas it was a net exporter of rice only a decade before.[9] This increasing dependency upon imported rice was seen as a potentially dangerous situation whereby Japan might end up relying upon foreign countries for its staple food supplies while expending valuable reserves of gold and foreign currency to pay for this trend. Given the nationalists' penchant for protection of domestic industries and emphasis on self-reliance, Meiji economic planners responded by trying to foster self-sufficiency in food production, most importantly in Japan's leading staple, rice. New policies, including a combination of subsidies, tariffs and the development of agricultural testing centres, were adopted in 1897 which aimed to boost the domestic production of rice while limiting food imports.[10] Stricter rice tariffs were subsequently applied in 1905 after significant lobbying efforts by farmers and their supporters.[11] Thus, Meiji economic planners were the first to associate the achievement of food security with the development of self-sufficiency.

Although rice was the most important food commodity to be targeted for support, it was not the only one. Meiji planners also attempted to foster development in the fisheries sector as part of its food self-sufficiency policy, with special emphasis on procuring a larger and more modern fishing fleet, supporting a domestic ship-building industry, and finding new fishing grounds abroad.

One of the first steps in this effort was enacting new fisheries legislation in 1902 which sought to commercialize fisheries by establishing associations that assigned fishing rights to coastal villages and marketed the products caught.[12] Also, several new fishing boats of superior British and Norwegian design were imported into Japan in the 1890s, which would later assist newly developed domestic shipbuilding industries construct their own ships with similar designs with the help of government subsidies after 1905.[13] Moreover, Japan's first refrigeration system was built in 1899 and ice-making facilities were pioneered later, which subsequently enabled fish caught at distant fishing grounds to be sold fresh in Japanese markets.[14] Coinciding with Japan's establishment of a protectorate in Korea and its annexation of the Takeshima Islands (Dokdo in Korean) in 1905, the Pelagic Fisheries Encouragement Act was enacted to promote distant-water fishing, especially near the Korean peninsula, and foster the construction of large, powered fishing vessels through subsidies.[15] The use of legislation, subsidies and promotion programmes demonstrate the determination of the Meiji government to foster a modernized fishing industry capable of meeting the domestic food needs of the nation.

With the exception of the brief experimentation of Taisho economic liberalism, subsequent economic policy making placed even greater emphasis on self-sufficiency in national planning. The rising influence of the military in domestic politics as well as the gradual expansion of the Japanese Empire in the early Shōwa era gave rise to an even stronger push for economic security in the form of imperial autarky. Michael Barnhart argues that Japan's efforts to develop a self-sufficient empire, one that could support a total 'economic war' if necessary, led to an aggressive, expansionist foreign policy eventually culminating in the Pacific War.[16] Autarky became a key feature of strategic planning in the 1930s which determined the course of subsequent economic policies in many sectors, including fisheries.

Fisheries in pre-war and Imperial Japan

Japan has a long tradition of coastal, beach and inland-water fishing that stretches back to well before 1900. It would be difficult to overstate the importance of fish in the diet of most Japanese and their consequently large role in the domestic economy. Fish products constituted more than half of the animal protein consumed in Japan, as opposed to about 5 percent in the United States and Canada.[17] The natural productivity of the waters surrounding Japan, the lack of alternative food sources and the concentration of population along the island's extensive coastlines led to the early intensive development of the local fishing industry.[18] Three distinct types of fisheries emerged in the early twentieth century distinguished both by their areas of operation and by their techniques: coastal, offshore and overseas.[19]

Overseas fisheries did not develop extensively prior to the mid-1920s as a result of the high cost and technical sophistication of refrigeration, diesel engines and larger ships. Even after the technology became available, only large companies such as Taiyo Gyogyō and Nippon Suisan (Nissui) were able to provide the capital investment and organizational expertise necessary to turn international expeditions into profitable ventures. Such fisheries primarily operated motorized factory ships and trawlers off the coasts of Russia, the Kwantung Peninsula, Korea, Formosa and the Japanese mandated islands in the mid-South Pacific, and mothership operations in the Sea of Okhotsk, the Bering Sea and the North Pacific high seas.[20] The main species exploited were sardines, tuna, skipjack, salmon, shrimp, crab and whales.

In terms of the relative importance of each fishery to the domestic Japanese economy in the pre-war period, coastal fisheries remained the largest supplier of fish. While offshore and overseas sources of supply continued to amount to only a fraction of the coastal catch in absolute terms, their production increased 7.5 times from the mid-1920s to 1940, and continued to provide commercially valued species such as whales, crab and salmon to the domestic market. Figure 1.1 does not consider colonial fisheries which supplied a yearly average of nearly 2.3 million tons of fish and whales – close to one-third of Japan's total production from 1935–40.[21] Since Korean coastal waters provided most of this total, one must not underestimate the importance to the domestic market of fishing products supplied from outside of Japanese coastal waters.

The increase in fisheries productivity was an important factor in the Meiji government's campaign of *fukoku kyohei* and in later governments' developmental strategies. Political leaders hoped that expanded fisheries would meet the dual functions of providing for the nutritional requirements of a rapidly growing population while simultaneously promoting enhanced economic activity in targeted industrial sectors. In 1905, the government enacted the Pelagic Fisheries Encouragement Act to assist the construction of large, motorized fishing vessels. Subsidies were also provided from 1918 onward to help the construction and repair of mechanized ships. Moreover, the government offered subsidies in 1923 and again in 1932 to promote refrigeration technology and ice-making facilities, respectively.[22] Such promotional policies greatly benefited those companies engaged in or expanding into overseas fisheries.

Japan's expanded fisheries productivity also had important environmental repercussions. Although general fishing effort in coastal and offshore trawling grounds substantially increased after 1933, catch rates actually levelled off.[23] This was an indication that the full utilization of marine resources was achieved by the early 1930s and that further effort merely resulted in overfishing.

Moreover, Japanese fishermen earned a poor international reputation for their over-exploitation of ocean resources. Japan refused to join the International Whaling Commission (IWC) in 1936, despite being the leading whaling country at that time, and the subsequent failure of this organization was blamed upon this kind of Japanese malfeasance.[24]

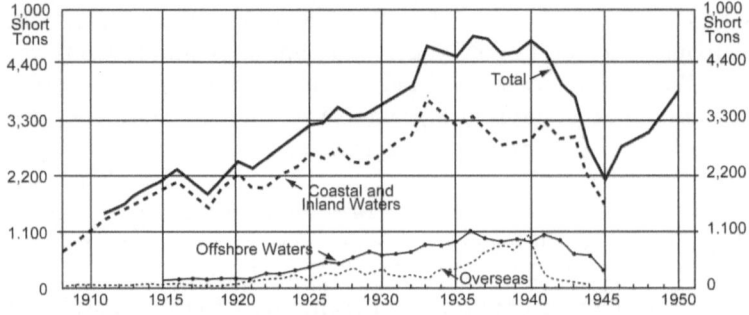

Figure 1.1 Natural resources section GHQ, SCAP, Tokyo
Source: Edward Ackerman, *Japan's Natural Resources and their Relation to Japan's Economic Future.* Chicago: Chicago University Press, 1953, 130

Not only was the failure of these protection measures attributed to Japan, but the decline of salmon stocks in the North Pacific was also correlated to the expansion of Japanese fishing efforts in the 1930s. Several exploratory salmon expeditions to Bristol Bay in 1936 also prompted the *Pacific Fisherman*, an influential North American industrial fisheries magazine, to warn of an impending 'alien invasion'.[25] Canadian and US fishermen feared that intensive Japanese fishing would produce unsustainable yields and possibly deplete this valuable North American stock. Thus, the expansion of Japan's fishing activity led in some cases to excessive exploitation of fisheries resources, and engendered international resentment against its fishing practices.

In step with the militarists' plans for autarky, colonial fish production came to supply a sizeable amount of seafood for the expanding Japanese home market, which was becoming more dependent upon intra-imperial trade. In the ten years between 1929 and 1938, colonial production averaged nearly 1.5 million metric tons a year, principally in waters near Korea, but also including smaller fisheries off Taiwan, Kwantung and the South Pacific mandated islands.[26] In 1938, fisheries based in Japan accounted for a total catch of 3.8 million metric tons, whereas colonial fisheries produced 2 million tons, or slightly less than half for the whole Japanese Empire.[27] Given Japan's direct control over neighbouring colonies, it would appear that self-sufficiency was achievable in the short run for many economic sectors, including fisheries food policy. Self-sufficiency in the long run, however, was unable to forestall Japan plunging into the disastrous Pacific War, which appeared to discredit immediately the strategic reasons given for the goal of self-sufficiency in the first place.

Conclusion

In facing the question of the *shigen mondai* and food security, Japanese policy makers have continuously opted for policies that promote self-sufficiency rather than relying upon foreign imports for its food needs. From early Meiji, policies that promoted domestic agricultural and fisheries production fostered a system of food controls and expanded fishing rights within the Empire. Proponents of self-sufficiency were successful in laying the foundation for food security thinking that exists in official circles in Japan today. In fact, these ideas are entrenched to the extent that they may be considered conventional wisdom for many of the elite.

National policies and programmes that fostered domestic production in an effort to achieve a high level of national self-sufficiency also created with it sectoral groups which had a vested interest in the preservation of the very system that led to their creation. In a sense, national interest as it was identified at critical junctures, such as the Meiji Restoration and the post-war Occupation, perpetuated national policy trajectories in a relationship that may be best described as path dependent.

This is most evident in the case of rice producers and distributors who have relied upon, and actively promoted the idea of, self-sufficiency as a desirable goal for food security. Indeed, the symbiotic relationship that rice producers and their nationwide organization, Nōkyō, have established with MAFF has subsequently ensured that self-sufficiency remained a hallmark of official planning and thinking in the bureaucracy throughout the post-war period. The same self-sufficiency strategy that has been designed for rice in the Food Agency has also been applied for fisheries policy in the closely related Fisheries Agency and, in this sense, food policy that has been adopted for rice production has guided policy for fisheries as well.

The prevailing regime of *mare liberum* in international waters facilitated the expansion of Japan's fishing activities in the early twentieth century. As steam- and then diesel-engine technology in conjunction with refrigeration facilities extended the range, catch capacity and transportation ability of Japan's fishing fleets, no legal framework or property rights existed to restrict such development. The unlimited entry nature of the common resource regime enabled growing Japanese fishing companies to expand their activities limitlessly into the international waters of the North Pacific and elsewhere to meet domestic consumption demands and export opportunities. Japan's offshore and distant water fisheries were dependent upon the 'freedom of the seas' concept in international practice for their very survival.

The following chapter will examine how Japan's quest for food self-sufficiency led to its wholehearted support of 'freedom of the seas' in international law and the concomitant expansion of distant-water fisheries in the early post-war era.

Notes

1 Douglas Johnson, *The International Law of Fisheries*. New Haven: Yale University Press, 1965, 165.
2 Larry Leonard, *International Regulation of Fisheries*. Washington, DC: Carnegie Endowment for International Peace, 1944, 7.
3 While neither principle was codified in international law, they were nonetheless regularly used in practice. As the international jurist Thomas Baty once observed with regard to the 3-mile limit: ' ... the rule, while not infrequently attacked in theory, is supreme in practice. Diplomatists seldom or never question it; professors occasionally do. In the actual conduct of affairs, it is seldom challenged, and never successfully so ... ' (Thomas Baty, 'The Three-Mile Limit', *American Journal of International Law* vol. 22 (July 1928): 503).
4 Richard James Sweeney, Robert Tollison and Thomas Willet, 'Market Failure, the Common-Pool Problem, and Ocean Resource Exploitation', *Journal of Law and Economics* vol. 17 (1974): 179–92.
5 Oran Young, *Resource Regimes: Natural Resources and Social Institutions*. Berkeley: University of California Press, 1982, 138.
6 Fred Sanderson, *Japan's Food Prospects and Policies*. Washington, DC: The Brookings Institution, 1978, 12.
7 Chuhei Sugiyama, *Origins of Economic Thought in Modern Japan*. London: Routledge, 1994, 2–12.
8 Ibid., 8.

9 Kym Anderson and Rod Tyers, 'Japanese Rice Policy in the Interwar Period: Some Consequences of Imperial Self Sufficiency', *Japan and the World Economy* vol. 4 (1992): 104.

10 Tōbata Seiichi, *Nihon nōgyō no sugata* (The Shape of Japanese Agriculture). Tokyo: Nihon Hyōronsha, 1953, 277.

11 Anderson and Tyers, 'Japanese Rice Policy', 104.

12 Yohoji Asada, Yutaka Hirasawa and Fukuzo Nagasaki, *Fishery Management in Japan*, FAO Fisheries Technical Paper 238. Rome: Food and Agriculture Organization, 1992, 6.

13 A. Niwa *et al.*, *Japanese Fisheries: Their Development and Present Status*. Tokyo: Asia Kyokai, 1957, 2.

14 *Taiyō Gyogyō 80-nen shi* (The Eighty-Year History of Taiyo Fisheries Corporation). Tokyo: Taiyō Gyogyō, 1960, 286.

15 Niwa *et al.*, *Japanese Fisheries*, 2.

16 Michael Barnhart, *Japan Prepares for Total War: The Search for Economic Security, 1919–1941*. Ithaca: Cornell University Press, 1987, 267–70.

17 Food and Agriculture Organization, *Relative Importance of Trade in Fishery Products*, 1973.

18 The warm Kuroshio current is largely responsible for providing the greatest source of diverse and plentiful fish species despite the relatively short fishing banks surrounding Japan.

19 Coastal fisheries were the oldest form of fishing in Japan. Such fisheries were largely conducted by village associations or diverse small-scale individual enterprises characterized by techniques requiring low capitalization, such as non-motorized boats, simple equipment and small-scale production. Their methods of operation included beach seines, lift nets, set nets or traps, gill nets and hook-and-line fishing within the customary 3-mile zone, as well as shellfish and seaweed collection, and inland fresh-water catches. The principal species landed were herring, salmon, trout, yellowtail, tuna, horse mackerel, bream, sharks, cuttlefish, octopus, crab, shrimp, clams and oysters. Offshore fishing, on the other hand, ranged from just outside the territorial sea to hundreds of miles from the coast. Due to the higher costs for the equipment and mechanized boats capable of operating at such ranges, offshore fishing was largely conducted by companies and special associations utilizing large-scale purse seine fishing, two-boat power trawling and line-and-pole tuna fishing techniques. The decentralized coastal fisheries with their low levels of capitalization were joined by new enterprise fisheries with much higher levels of capitalization as fleets extended into offshore areas. Such fisheries were largely based on tuna, skipjack, sardines, mackerel, cod, bream, sharks, flatfish, skipper and mullet. Edward Ackerman, *Japan's Natural Resources and Their Relation to Japan's Economic Future*. Chicago: Chicago University Press, 1953.

20 Ibid., 115.

21 Ada Espenshade, *Japanese Fisheries Production, 1908–46*. SCAP Natural Resources Section Report 95, found in Ackerman, *Japan's Natural Resources* (in the original 1948 version), 132.

22 Niwa *et al.*, *Japanese Fisheries*, 2.

23 Ackerman, *Japan's Natural Resources*, 444.

24 Douglas Johnson, *The International Law of Fisheries*. New Haven, CT: Yale University Press, 1965, 396–411.

25 *Pacific Fisherman* vol. 34, no. 10 (October 1936).

26 Ackerman, *Japan's Natural Resources*, 115.

27 Ibid., 132. No imports were recorded during this time.

Bibliography

Ackerman, Edward. *Japan's Natural Resources and their Relation to Japan's Economic Future*. Chicago: Chicago University Press, 1953.

Anderson, Kym and Rod Tyers. 'Japanese Rice Policy in the Interwar Period: Some Consequences of Imperial Self Sufficiency', *Japan and the World Economy* vol. 4, no. 2 (September 1992): 103–27.

Asada, Yohoji, Yutaka Hirasawa and Fukuzo Nagasaki. *Fishery Management in Japan*. FAO Fisheries Technical Paper 238. Rome: Food and Agricultural Organization, 1992.

Barnhart, Michael. *Japan Prepares for Total War: The Search for Economic Security, 1919–1941*. Ithaca: Cornell University Press, 1987.

Johnson, Douglas. *The International Law of Fisheries*. New Haven, CT: Yale University Press, 1965.

Leonard, Larry. *International Regulation of Fisheries*. Washington, DC: Carnegie Endowment for International Peace, 1944.

n.a. *Taiyō Gyogyō 80-nen shi* (The Eighty-Year History of Taiyo Fisheries Corporation). Tokyo: Taiyō Gyogyō, 1960.

Niwa, A. *et al.*, *Japanese Fisheries: Their Development and Present Status*. Tokyo: Asia Kyokai, 1957.

Sanderson, Fred. *Japan's Food Prospects and Policies*. Washington: The Brookings Institution, 1978.

Sugiyama, Chuhei. *Origins of Economic Thought in Modern Japan*. London: Routledge, 1994.

Sweeney, Richard James, Robert Tollison and Thomas Willet. 'Market Failure, the Common-Pool Problem, and Ocean Resource Exploitation', *Journal of Law and Economics* vol. 17 (1974): 179–92.

Tōbata Seiichi. *Nihon nōgyō no sugata* (The Shape of Japanese Agriculture). Tokyo: Nihon Hyōronsha, 1953.

United Nations Food and Agricultural Organization. *Relative Importance of Trade in Fishery Products*. Rome: FAO, 1973.

Young, Oran. *Resource Regimes: Natural Resources and Social Institutions*. Berkeley: University of California Press, 1982.

2 Ocean regimes and food system planning under SCAP occupation

This chapter will look at the situation of ocean law and policy, food security and fisheries in the immediate post-war period (1946–52). It will start by explaining the conditions of the fishing industry in the immediate post-war era and the challenges posed to fisheries interests during the Occupation of Japan. An examination of how Japan navigated its way through various fishery regime alternatives, including the expanded jurisdictions claimed by the Truman Proclamation, and how the United States assisted Japan by establishing favourable international precedents will reveal the prudence and effectiveness with which the Japanese were able to use regional power balances to their advantage and to avoid the ever-present possibility of fisheries restrictions. In the following chapter, it will be shown how Japan's motive to secure a high degree of food security through self-sufficiency was largely achieved as a result of the liberal fisheries agreements negotiated during this time.

The early post-war environment[1]

The Pacific War had devastated Japan's fishing capacity. Many fishing boats requisitioned for war service were sunk or heavily damaged, to the extent that only 40 percent of the 1940 total of powered fishing-boat tonnage was still operational after the war.[2] Many port facilities also suffered extensive bombing damage so that expensive and time-consuming repairs were required before they could become serviceable. As a consequence, the fish catch dropped by half in the immediate post-war period while consumption fell by one-third from the pre-war peak in 1937.[3] The Supreme Commander for the Allied Powers (SCAP) faced a dire situation of widespread undernourishment threatening Japan upon arriving in Tokyo in August 1945.[4]

The new Occupation authorities in SCAP and the Far Eastern Commission (FEC) first engaged in heated debates with regard to what kind of Japan to rebuild and how to do it. In general, two alternatives were considered with respect to Japan's economic recovery: restriction and revival. Several Allied powers, including Australia, the USSR and Great Britain, favoured a policy of restriction whereby the Japanese economy would be permitted to rebuild to the level of self-sufficiency but not to the point where renewed prosperity might once again compromise regional security. In the Occupation's early stage (1945–47), the United States tempered such calls for restriction but it also did not unreservedly support the renewal of the Japanese economy.[5] This was a time when Washington and SCAP were trying to define a new role for Japan in the post-war order.

Fisheries were understood to play an important role in the Japanese economy and were hence deemed critical for economic revitalization. SCAP policy with respect to Japan's fisheries fluctuated between the alternatives of restriction and revival, as can be seen through the differing fisheries regime options it was entertaining at the time. Washington's consideration of the type of fisheries regime to apply to Japan parallels the options considered for many other areas of the Japanese economy and provides an interesting insight into shifting attitudes of US planners in the post-war world.

The Truman Proclamation

As seen in Chapter 1, Japan's offshore and distant-water fisheries were dependent upon the 'freedom of the seas' concept in international practice for their very survival. Would Japan be able to defend this regime alternative, given its position as a nation vanquished in the Pacific War?

In September 1945, shortly after the defeat of Japan, the United States made a significant challenge to the *mare liberum* system with its announcement of the Truman Proclamation. The text of the unilateral declaration read as follows:

> In view of the pressing need for conservation and protection of fishery resources, the Government of the United States regards it as proper to establish conservation zones in those areas of the high seas contiguous to the coasts of the United States wherein fishing activities have been or in the future may be developed and maintained on a substantial scale.[6]

In essence, the United States laid claim to fisheries and other resources well into international waters, which in the case of the American continental shelf extended in places to more than 200 nm. While the United States took great pains to reassure the international community that this proclamation did not equate to a declaration of sovereignty beyond its 3-nm territorial sea, it was the first major challenge to prevailing international law and it entailed, in particular, significant repercussions for Japanese distant-water fisheries.

The Truman Proclamation challenged the *mare liberum* concept by delineating a much extended limit to the control of fisheries and other resources into the open ocean. At a time when the United States was setting new international rules and regimes with regard to international finance, shipping and air traffic control, among others, the declaration in effect proposed a post-war oceans system of coastal state appropriation of contiguous fishery resources that would have mirrored the exclusive economic zone (EEZ) system later instituted by the United States in 1976 and validated by United Nations Convention on the Law of the Sea (UNCLOS) in 1982.[7] The Truman Proclamation threatened severely to impede Japanese distant-water fisheries.

President Truman was motivated to make this proclamation in an effort both to protect American tuna fisheries in the Gulf of Mexico and to exercise greater control of Pacific-coast salmon in the Bering Sea. The Alaskan fishing industry feared a revival of Japanese salmon fishing in Bristol Bay and lobbied to pre-empt this with what essentially served as a declaration of ownership of the salmon resource.[8] While this alone would not have been fatal to Japanese fishers, the proposed change in resource regimes would have been calamitous if other countries in the North Pacific and elsewhere followed suit by declaring similar zones over other fishing grounds. The most significant feature about the declaration is that it served as an alternative post-war regime to freedom of the high seas – one that severely threatened Japanese fishing interests.[9]

Contention among the Allies and fisheries restrictions

At the outset of the Occupation of Japan, the United States was clearly contemplating various alternatives with respect to a new oceans regime in the Pacific and elsewhere. As new international priorities arose during the course of the Occupation with the onset of the Cold War, Washington began to reconstitute its relationship with Japan through a reverse course in economic and political planning.[10] As a consequence, the United States realigned its oceans policy in favour of freedom of the seas. This helped restore Japan as the world's greatest fishing nation.

The initial fisheries policy for Japan was drafted by the State-War-Navy Coordinating Committee (SWNCC) in August 1945 and its provisions were used to guide SCAP authorities in the early years of the Occupation. The policy stated that the Occupation authorities would enable the rapid restructuring of fisheries capacity to produce the required food for Japan in view of widespread starvation.[11] SCAP would also permit fishing in the deep sea and near the coasts of other nations when domestic fisheries were unable to fulfil domestic requirements.[12]

Soon after Japan's surrender, SCAP had suspended the movements of all vessels weighing over 100 tons in the interest of security, but such restrictions were now relaxed.[13] Motor-powered fishing vessels were exempted on 27 September 1945 within an authorized zone known as the MacArthur Line (see Figure 2.1), in accordance with the SWNCC directive.[14] For the next six years, Japanese fishing operations were limited to Japanese coastal waters within this fishing boundary – only a fraction the size of pre-war fishing areas. In the early years of the Occupation, SCAP authorities sought to restrict the activities of Japanese fishermen and began a reassessment of the need for reviving distant-water fleets altogether.

Despite the strict limitations placed upon Japanese fishing in the early Occupation period, many Allied members of the Far Eastern Commission were still alarmed at the prospect of a renewed Japanese fishing industry and sought to place even greater limitations upon their activities. George Blakeslee, a contemporary political adviser to the FEC chairman, characterized the ensuing fisheries discussions among the Allies as engendering more hard feelings and acerbic discussions than any other topic deliberated in the commission.[15] Australia was the most outspoken proponent of the 'containment' of Japanese fisheries on a permanent basis, although the Australians initially had strong support from New Zealand, the United Kingdom, China, the Philippines, the Netherlands and Norway.[16]

At the same time as imposing restrictions on fishing activities, however, SCAP authorities also recognized the need to avert a food crisis in Japan and were keen to defray some of the costs of providing food aid to this impoverished nation. Such motives would set the stage for post-war fisheries policy that emphasized food security for Japan. In perhaps one of the most hotly contested decisions in the FEC, SCAP authorities unilaterally overrode tremendous opposition among the Allies to reopening Japan's Antarctic whaling

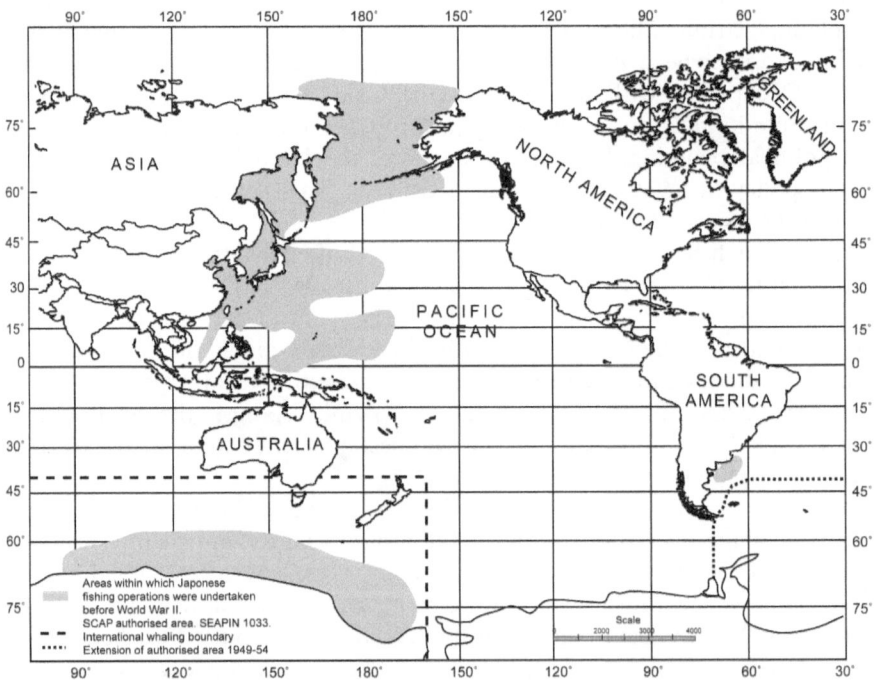

Figure 2.1 Japanese pre-war fishing areas (dark areas) and limits of authorized fishing areas (lines), 1945–51

Source: Edward Ackerman, *Japan's Natural Resources and their Relation to Japan's Economic Future.* Chicago: Chicago University Press, 1953

operations in the summer of 1946, arguing that Japan needed the protein for the nation's food supply and that the operation would be conducted 'one time only'.[17] Whaling expeditions to the Antarctic, however, were to continue every subsequent year of the Occupation, much to the chagrin of other FEC member nations.

The reverse course

By 1947–48, the United States began to see its relationship with Japan in new Cold War terms and embarked on a 'reverse course', modifying some of the initial economic and political reforms for Japan, including plans for Japan's international fisheries. As Washington grew increasingly wary of communist encroachments in East Asia and Europe, SCAP began to support the idea of reviving Japan as an economically strong power to serve as an anti-communist bulwark in Asia.[18] The United States used its pre-eminent position in SCAP, the Far Eastern Commission and the Allied Council on Japan (ACJ) to promote the renewal of Japan's economic power.[19]

SCAP authorities soon identified the fishing industry as an area critical to economic revitalization. It was hoped that a strong fishing industry would allow the Japanese to provide for their own food requirements while relieving the United States of burdensome aid expenses and create the necessary impetus to rebuild essential economic sectors such as ironworks and ship-building.[20] Furthermore, exports of surplus fish products could provide much needed hard currency and help build foreign exchange reserves. Thus, fisheries were promoted to alleviate the nation's post-war economic crisis, including serious food shortages, and to foster economic recovery. As will be seen, these motives served as the cornerstone of Japanese fisheries policy in the post-war period.

Whereas initial discussions on Japanese fisheries centred on restrictions and a punitive settlement in the peace treaty, SCAP used the reverse course to ensure a non-vindictive treatment of fisheries that emphasized their expansion. SCAP and the State Department agreed that Japan should be protected from Allies seeking to limit its access to world resources, including fisheries, and from efforts to burden Japan with reparations or other financial penalties that might hamper its redevelopment.[21]

For its part, Japan used the reverse course to push for an expansion of fishing areas and extended terms for whaling. In response to Australian and other Allied objections to furthering the whaling effort in the Antarctic and calls for greater restrictions on Japan's high seas fishing, State Department officials in the International Resources Division warned that any concessions to the Allies might be construed as a precedent and would amount to a capitulation to the philosophy of fisheries restrictions in the future, particu-larly during Peace Treaty negotiations.[22] In September 1949, SCAP author-ized the third eastward expansion of the MacArthur Line and in May 1950, gave permission for tuna long-liners to operate in the Trust Territories – both decisions proved very controversial among the Allies, particularly with Australia.[23]

The Chief of SCAP's Natural Resources Section (NRS) Fishery Division, William Herrington, was also strongly opposed to Allied efforts to impose restrictions upon Japanese high seas fishing. As the chief negotiator for the United States during trilateral fisheries talks with Japan in 1952, Herrington wished to ensure that no precedent was set for other nations to ban Japanese fishing operations from their coasts, especially when these fisheries were not under conservationist management.[24] Instead, he tried to have Japanese officials support the notion of the scientific management of fisheries and to demonstrate that Japan was an environmentally responsible nation.[25]

The United States also reversed the course of the Truman Proclamation. Washington chose not to enforce the Proclamation's provisions and it was allowed to lapse in practice, although it still served as a precedent for other countries' coastal claims in later years. Perhaps fearing that extended coastal claims might set an adverse precedent for Japan elsewhere in the North Pacific, the State Department disassociated itself from the Truman Proclamation by publishing two papers in December 1948 and April 1949 reasserting its adherence to the 3-mile rule.[26] Moreover, while the Truman Proclamation served as the first instance of an extended coastal claim over fisheries, the United States vehemently opposed similar claims by other states in subsequent years. This extended jurisdiction declaration would not be reactivated for another thirty years.

Revival of fishing industries

Another vital component of fisheries revival was the rebuilding of fishing fleets. The war did not have as destructive an impact on shipbuilding yards as it did on fishing fleets and docks, which may explain why the shipping industry was able to stage a rapid recovery in the early post-war years.[27] Such a recovery was also partly attributable to the various policies promoted by the SCAP and Japanese administrations, such as subsidies and grant-matching programmes including the Programmed Shipbuilding Scheme.[28] Although shipping was carefully regulated and various restrictions were applied to the shipbuilding industry in the early phase of the Occupation, these policies were eventually reversed or simply not applied in the later years given the new anti-communist orientation of the United States. Consequently, Japan was able to rebuild its fishing fleets and shipping liners to an extent that the shipbuilding industry achieved the world's top position within a mere decade after the war.[29] This naturally gave a boost to the fishing industry by providing a ready, subsidized supply of vessels for the recapitalization of the fisheries sector.

These promotional policies were successful to the extent that by 1947 the restored fishing-fleet tonnage exceeded pre-war levels. In spite of considerable population growth, indigenous production of agriculture and fisheries was sufficient to provide 66 percent of Japan's calorific requirements – close to pre-war levels. Yet Japan still suffered from a food deficit of 11 percent even as late as 1950, according to SCAP studies.[30] It was soon recognized that in order to allow Japan to continue to develop the full capacity to sustain its own economy, SCAP needed to extend or find new fishing areas for Japanese fishers.

To assist in extending its fisheries activities, Japan tried to alleviate international fears that it might once again ruthlessly over-exploit common fisheries resources and made efforts to portray itself as an internationally responsible nation. Most of the worldwide opposition to Japanese distant-water fisheries – including American, Canadian and Australian fishing industries – stemmed from Japan's previous lack of proper conservation practices.[31] In 1950, Prime Minister Yoshida approached US Ambassador John Foster Dulles and the missions of various governments in Tokyo with an official declaration announcing Japan's willingness to abide by worldwide conservation standards and agreements regarding ocean resources.[32] Accordingly, Japan joined the International Convention on the Regulation of Whaling (ICRW) in April 1951.[33] Furthermore, the threat of overfishing in the authorized area also prompted Tokyo to make better use of available resources through more rigorous conservation measures beginning in 1950.[34] Japan sought to expand its fishing activities by voluntarily adopting international environmental standards, which provided a reasonable justification for SCAP to permit an extended fisheries area for Japan.

By 1952, it was clear that SCAP officials wished to promote a revival of Japan's distant-water fisheries. Japanese bureaucrats and politicians, too, sought to increase the fisheries catch in order to provide food security to a nation that had only recently experienced food shortages and under-nourishment. How did thinking on food security develop during the Occupation and what were the motives behind it?

Food security and post-war economic planning

The United States-led administration in charge of resource planning, the Natural Resources Section (NRS), devoted considerable energies to solving the question of how Japan could overcome the problem of inherent resource scarcity, or the *shigen mondai*, especially in light of drastic food shortages shortly after the war and Japan's post-war international isolation. Japanese economic planners worked with the SCAP administration to devise strategies that could address short-term food scarcity and longer-term food security for the nation, and even created some of their own independent recommendations for resource planning for the economy.

The Natural Resources Section conducted a plethora of studies on resources on the Japanese islands and in the surrounding seas, especially focusing on materials, lands and natural resources that could be harnessed and used in the reconstruction of the nation. These studies were later used to help shape resource allocation planning for the Occupation and Japanese governments. The NRS completed a total of 230 reports and preliminary studies relating to all natural resources sectors in Japan, representing the largest comprehensive study of resources undertaken on Japan to that date.[35] Of these, thirty-one reports and studies were conducted on Japanese fisheries as prepared by the Fisheries Division of the NRS.

The most important of these SCAP studies was a comprehensive report completed by Edward Ackerman between 1948 and 1949, and later republished with some revisions and expansions in 1953 under the title of *Japan's Natural Resources and their Relation to Japan's Economic Future*.[36] The report examines Japan's resources and future requirements in terms of food, energy, fibre, and non-mineral and mineral supplies for construction and industry. Many of the report's analyses and recommendations were used by SCAP and later Japanese officials in developing resource allocation policies in the post-war era, and thus can be said to have exerted an influence on official thinking in the Japanese government. Government planning in the economy was assumed to be not only desirable, but necessary for the reconstruction of the economy.

Ackerman's report clearly stated near the outset that among Japanese needs, food took precedence over all others. In order to solve the food shortage problem in the long run, Japan had three options given its constantly growing population: to increase food imports, or increase food production at home, or both.[37] Ackerman clearly favoured increasing domestic food production in his recommendations and thereby supported a guiding policy of self-sufficiency. He noted that Japan nearly reached the peak in food production possibilities in the interwar period due to the very intensive development of fisheries, but also suggested that there still existed several possibilities for increased indigenous food production in agriculture if cultivation was expanded and improved upon, fertilizers were used and food-crop pests were controlled.[38] Perhaps owing to Japan's international isolation at the time, Ackerman did not invest much faith in increasing imports in the short term. Instead, his recommendations were based on the thinking that Japan must be capable of meeting its food needs on its own: 'Japan can hope to be adequately or well nourished only when it raises domestic production substantially ... '[39]

Ackerman thought that fisheries could make an important contribution to Japan's food security problem and recommended an expanded catch to meet domestic food needs. He noted that in view of the fact that fisheries products had supplied 80–90 percent of all animal protein consumed in Japan and the daily per capita intake was 17.5 grams, Japan would need to at least double its 1947 fisheries catch by the year 1965.[40] However, he stated that the limit of productivity of coastal waters had already probably been exceeded.[41] Thus, the elimination of restrictions on Japanese fishing and the consequent development of a distant-water fishing capacity was the most likely solution to assist in creating a self-sufficient food supply.

It was not only SCAP studies that pondered the 'resource problem' and food security in post-war economic planning. Several Japanese officials in the government and in government-affiliated research institutes published research reports that had an influence on official government thinking in post-war economic planning. Two such reports were *The Postwar Reconstruction of the Japanese Economy*, which was compiled by Okita Saburo of the Ministry of Foreign Affairs in September 1946, and *Nihon no Shigen Mondai* (Japan's Resource Problem), which was published by the Economic Stabilization Board (ESB) in 1952.[42] While neither report was officially adopted as government policy, both were widely read in government circles and made significant contributions to the debate on Japan's economic future.

Okita Saburo's report was originally published as a special survey commission sponsored by the Ministry of Foreign Affairs, but the Ministries of Finance, Agriculture and Forestry, and Commerce and Industry also cooperated in its drafting.[43] It is probably best known as the study that provided the conceptual basis for the adoption of the priority production system, which was centred on increasing coal production, and the economic rehabilitation plans drafted subsequently. Although the report proposals were not necessarily reflected immediately and directly in the administration of the economy, it provided vision to the Japanese who read it at the time and exerted an indirect influence on the management of the Japanese economy thereafter.[44]

Although most of the report is devoted to re-industrialization of the economy through priority production of heavy industries, it also addresses other concerns for a rehabilitated Japanese economy, including food security. The report rules out the possibility of increasing food imports into Japan, noting that 'successive imports of large quantities of food year after year will impose a heavy burden on the nation's international payments, will additionally make difficult the import of industrial raw materials and other essential goods, and will impose a constraint on the industrialization of the country'.[45] Thus, self-sufficiency in food production was lauded as the best solution to Japan's problems of food security and industrialization, and therefore must 'be attained to the maximum degree to the extent that it will not result in an extreme waste of labor'.[46]

Fisheries were also featured in *The Postwar Reconstruction of the Japanese Economy* as one of the specific challenges facing the Japanese economy at the time. Noting the importance of fisheries products in protein supply of the nation, the report recommended that fisheries 'be revived to the maximum extent', and that distant-water fishing be promoted 'at the earliest possible date'.[47] For the revival of deep sea fishing to take place, the report added, Japan must win a 'favourable understanding of the situation by the Allied Powers'.[48] The section on fisheries concluded by saying, 'the aptitude of the Japanese as fishermen and their high standards of fishing techniques should be used to the utmost for establishing fishery industries throughout the world. Japanese fishermen and fishing technology are expected to advance into fishing grounds throughout the globe in the future'.[49] Thus fisheries were seen as both a means of feeding the Japanese nation and as an industry with significant future economic potential.

A second study of considerable importance to post-war planning was the ESB's *Nihon no shigen mondai*, published in 1952.[50] This study was authored by Aki Kōichi, a professor at the University of Tokyo and chairman of the ESB's Natural Resources Inspectorate, and was inspired by Ackerman's SCAP report on Japan's natural resources. It sought to address and update some of Ackerman's findings and recommendations.

Nihon no shigen mondai examined Japan's food situation in the immediate post-war era and offered several recommendations to ensure stable production in future years. One of the report's conclusions was that unless more arable land was found, new modern techniques and technology employed on farms, and support for the agricultural sector offered by the government, Japan would become even more dependent upon imports to meet its food requirements.[51] It recommended fostering an agricultural industry capable of meeting domestic food needs while dismissing imports as a possible long-term solution to Japan's food shortages.[52] Aki deemed self-sufficiency desirable for food security over a reliance on foreign markets.

Domestic food system

In addition to such post-war economic studies and policy planning, a legal framework also existed that perpetuated the goal of self-sufficiency in food security planning. Most importantly, the Food Control Law of 1942 regulated and controlled Japan's domestic food market, especially for rice, through controls on price, distribution and trade until it was replaced by a new law with similar goals in November 1995. While the food control system did not target fisheries per se, the goal of self-sufficiency upon which the system was predicated became a central tenet for bureaucrats in the Ministry of Agriculture, Forestry and Fisheries (MAFF), and thus had an impact in their thinking on questions relating to food security and fisheries in subsequent years. Rice policy during this time had the effect of guiding overall food security policy in general.

Japan's wartime food shortages compelled the government to set up a system by which food was produced, priced and distributed. The Food Control Law of 1942 was designed to ensure the equitable distribution of rice and other grains under a situation of scarce supply and government control of the economy during wartime. It created two public utilities, the Central Food Corporation (*Chūō shokuryō eidan*) and the Regional Food Corporation (*Chihō shokuryō eidan*) which purchased rice, stored and processed it, and then distributed it and other cereals via distribution stations that they directly controlled.[53] Article 11 of the Food Control Law also banned all imports of rice, seeking to make Japan self-sufficient in production during wartime.[54] Under this system, almost all staple commodities fell under a rationing system controlled by the government which aimed to create an independent food supply.

After the defeat of Japan in 1945, the food crisis continued and the food control system remained in place. As the threat of serious food shortages lifted in 1947, in part due to aid and food imports offered by the United States, some aspects of food control relating to consumer rationing were liberalized. Government regulation of rice collection and distribution, however, remained an entrenched feature of the system for decades to come.[55] As Erich Pauer noted in his study of the Japanese war economy, Japan's surrender did not necessarily bring the economy to a standstill nor result in immediate changes: 'there was still a Japanese government headed by a prime minister, the whole administration network remained almost unchanged for a while, and the whole basket of laws and ordinances governing the planned economy changed only slowly'.[56] The Food Control Law continued to govern staple food production and distribution even in the post-war era since one of the foremost considerations of the Japanese government in the aftermath of the devastating Pacific War was to ensure a dependable food supply, especially principal grains, through the promotion of domestic agriculture production.[57] In a sense, self-sufficiency and its legal manifestation, the Food Control Law, served to justify continued government intervention in the economy in order to foster rural reconstruction and subsidize domestic agriculture.

Thus the legal requirement for self-sufficiency that was designed for supporting domestic rice production became one of the organizing philosophies of MAFF and, in particular, its subsidiary Food Agency (Shokuryō-chō) in the post-war period. As Japan readied itself for self-government in 1951–52, MAFF coordinated with other government agencies to undertake a three-year plan for 'economic independence' (*keizai jiritsu keikaku*) that called for the establishment of a long-term financial system, investment guarantees for energy, shipbuilding and agricultural industries, as well as increasing the rate for self-sufficiency, among other policies.[58]

Although policies for increasing self-sufficiency largely focused on agricultural reforms and the creation of agricultural cooperatives, MAFF and the Fisheries Agency also targeted expanded fishing rights as a means of securing necessary food reserves. Part of this expansion was achieved through the creation of fisheries cooperatives in coastal communities to which the government sold and allocated coastal and offshore fishing licences in an effort to 'privatize' (*mineika*) the industry.[59] Since the newly established Basic Fisheries Law of 1949 also contained provisions for distant-water fishing licensing as well, these too would play an important role in expanding fishing rights and capacity in an effort to increase food production.[60] Thus MAFF's goal to achieve a significant measure of self-sufficiency in food production influenced the aims and policies of the Fisheries Agency too.

Now that self-sufficiency was reaffirmed as a goal of national interest, Japanese officials set about negotiating access agreements for its distant-water fisheries abroad. Taking advantage of the preponderant position of the

United States in Pacific-area relations, Japan sought to conclude fishing treaties with neighbouring countries as stipulated in Article 9 of the San Francisco Peace Treaty. How negotiations were conducted, what diplomatic positions Japanese delegations took in various bilateral negotiations, and why the resulting agreements undertook specific normative and procedural stances are the subjects of the next chapter.

Notes

1 The latter part of this chapter was previously published in Roger D. Smith, 'Food Security and International Fisheries Policy in Japan's Postwar Planning', *Social Science Japan Journal* vol. 11, no. 2 (2008): 259–76.
2 A. Niwa *et al.*, *Japanese Fisheries: Their Development and Present Status.* Tokyo: Asia Kyōkai, 1957, 2.
3 Ibid., 2–3.
4 The Supreme Commander for the Allied Powers (SCAP) is used to refer to both the commander of the Occupation of Japan, Douglas MacArthur, as well as the official Occupation, its polices and administration.
5 Michael Schaller, *The American Occupation of Japan: The Origins of the Cold War in Asia.* Oxford: Oxford University Press, 1985, 20–52.
6 Charles B. Selak, Jr, 'Recent Developments in High Seas Fisheries Jurisdiction Under the Presidential Proclamation of 1945', *American Journal of International Law* vol. 44 (October 1950): 670.
7 For examples of the United States' role in the creation of other regimes, see Mark W. Zacher and Brent Sutton, *Governing Global Networks: International Regimes for Transportation and Communications.* Cambridge: Cambridge University Press, 1996. Indeed, the influence of the United States was such that the proclamation served as a precedent for similar declarations made by several Latin American countries, namely Peru, Argentina and Chile in 1952, and by many other countries later.
8 Kawakami Kenzō, *Sengō no kokusai gyogyō seido* (The Post-war International Fisheries Regime). Tokyo: Dai Nihon Suisankai, 1972, 107.
9 This may have been used as a negotiating ploy during the Occupation period to extract larger concessions from Japan, which was eager to avert the establishment of a highly restrictive fisheries system. Despite the significant threat this regime posed to its fishing interests, Japan did not register a formal objection with the United States, probably owing to Japan's recent status as a vanquished nation under occupation. It is possible that some kind of informal objection was made at the personal level. The Japanese government at this time was still in the first stages of reconstituting itself and the Fisheries Agency (as part of the Ministry of Agriculture and Forests) was not established until 1948.
10 The reverse course was a change in US Occupation policy regarding Japan from demilitarization, democratization and anti-trust policies to a focus on anti-communism and bringing Japan firmly onto the US side of the Cold War. Schaller, *The American Occupation of Japan*, passim.
11 US Department of State, *Foreign Relations of the United States (FRUS)*, 1945, vol. I, 612.
12 Ibid., 612.
13 Niwa *et al.*, *Japanese Fisheries*, 3.
14 Kawakami, *Sengō no kokusai gyogyō seido*, 16–22.
15 George Blakeslee, *The Far Eastern Commission: A Study in International Cooperation, 1945–1952*, Far Eastern Series 60. Washington, DC: Department of State, 1953, 105.

16 *FRUS*, 1947, vol. VI, 475–76.
17 Blakeslee, *The Far Eastern Commission*, 106–14.
18 William S. Borden, *The Pacific Alliance: United States Foreign Economic Policy and Japanese Trade Recovery, 1947–1955*. Wisconsin: University of Wisconsin Press, 1984, 81–82; Roger Buckley, *Occupation Diplomacy: Britain, the United States and Japan, 1945–1952*. Cambridge: Cambridge University Press, 1982, 126–27; Theodore Cohen, *Remaking Japan: The American Occupation as New Deal*, Herbert Passin, ed. New York: Free Press, 1987, 146–53.
19 Henry Esterly, 'Overseas Fisheries and International Politics in the Occupation of Japan, 1945–52', *The Occupation of Japan: Economic Policy and Reform*, Lawrence Redford, ed. Norfolk: MacArthur Memorial, 1980, 91–123.
20 In 1949, SCAP spent 45 percent of its US$515 million relief funds on food. Esterly, 'Overseas Fisheries', 94.
21 Schaller, *The American Occupation of Japan*, 122–40.
22 *FRUS*, 1947, vol. VI, 179–80.
23 Kawakami, *Sengō no kokusai gyogyō seido*, 23–30. The Trust Territories were a group of formerly occupied southern islands that extended as far south as Saipan and Guam.
24 William Herrington, 'In the Realm of Diplomacy and Fish', *Ecology Law Quarterly* vol. 16 (1989): 102.
25 Harry Scheiber, *Inter-Allied Conflicts and International Law, 1945–53: The Occupation Command's Revival of Japanese Whaling and Marine Fisheries*. Taipei: Academia Sinica, 2001, 64.
26 Harry Scheiber, 'Pacific Ocean Resources, Science, and Law of the Sea: Wilbert M. Chapman and the Pacific Fisheries, 1945–70', *Ecology Law Quarterly* vol. 13 (1986): 457–64.
27 Tomohei Chida and Peter Davies, *The Japanese Shipping and Shipbuilding Industries: A History of their Modern Growth*. London: The Athlone Press, 1990, 12.
28 Ibid., 24.
29 Ibid., 57–59.
30 This figure is based upon an approximate 1,900 calories/day consumption for an adult requirement of 2,250 calories/day. Edward Ackerman, *Japan's Natural Resources and their Relation to Japan's Economic Future*. Chicago: Chicago University Press, 1953, 153, 160.
31 Scheiber, *Inter-Allied Conflicts*, 175–95. Some examples include the Bristol Bay controversy and Australia's objections to renewed Japanese fishing.
32 Ministry of Foreign Affairs (MOFA), *Kita taiheiyō no kōkaigyogyō ni kansuru kokusai jōyaku kankei iken* (Opinions Regarding International Agreements on National Fisheries in the North Pacific). Microfilm Series B'0039, MOFA Archives.
33 Japan Whaling Association, www.whaling.jp/english/history.html.
34 Japan's first fisheries conservation law, the Law for the Prevention of the Exhaustion of Marine Resources, was enacted on 1 May 1950 to restrict catch rates within the MacArthur Line. The law was primarily designed as an ad hoc arrangement to mitigate an impending ecological crisis in the authorized zone. Fisheries Agency, *Suisanchō 50-nenshi* (Fifty-Year History of the Fisheries Agency). Tokyo: Suisanchō 50-nenshi kyōkai, 1998, 95.
35 Ackerman, *Japan's Natural Resources*, appendices VII and VIII.
36 Ackerman was a professor of geography at the University of Chicago who was hired as a technical adviser and visiting expert consultant for SCAP.
37 Ackerman, *Japan's Natural Resources*, 376.
38 Ibid., 376.
39 Ibid., 455.
40 Ibid., 450–51.
41 Ibid., 444.

42 Okita Saburo, *Postwar Reconstruction of the Japanese Economy*. Tokyo: University of Tokyo Press, 1992; and Aki Kōichi, *Nihon no shigen mondai* (Japan's Resource Problem). Tokyo: Kokon Shoin, 1952.
43 Okita, *Postwar Reconstruction of the Japanese Economy*, xvi.
44 Ibid., xxviii.
45 Ibid., 174.
46 Ibid., 174. The reference to 'extreme waste of labor' may be a qualification acknowledging that pre-war goals of self-sufficiency might have 'gone too far' by ignoring the heavy costs in terms of labour, economic efficiency and, ultimately, peaceful foreign relations in the pursuit of this goal.
47 Ibid., 145.
48 Ibid., 146.
49 Ibid., 146.
50 Aki, *Nihon no shigen mondai*. The ESB was established under the control of the prime minister in 1946 and it sought to assist in 'making emergent policies for economic stability relating to the production, distribution and consumption of goods, services, prices, finance, transportation and others, and to do the office work for the coordination, inspection and promotion of stability' (Article 1 of the Ordinance of the Economic Stabilization Board, found in Aiko Ikeo, 'Economists and Economic Policies', in *Japanese Economics and Economists Since 1945*, Aiko Ikeo, ed. London: Routledge, 2000, 147). The ESB is widely recognized as having exercised strong influence over the development of economic planning in the post-war economy.
51 Aki, *Nihon no shigen mondai*, 35–36.
52 It is interesting to note that *Nihon no shigen mondai* did not explicitly address the question of fisheries' contribution to Japan's food needs in the post-war period. The author states in his preface that several topics still lacked adequate statistics and printed sources, which resulted in an incomplete report of the full situation of Japan's resources. This may have been the case with the subject of fisheries, especially since later publications by the same author addressed fisheries resources directly.
53 Erich Pauer, 'A New Order for Japanese Society: Planned Economy, Neighbourhood Associations and Food Distribution in Japanese Cities in the Second World War', *Japan's War Economy*, Erich Pauer, ed. London: Routledge, 1999, 90.
54 Aurelia George Mulgan, *The Politics of Agriculture in Japan*. London: Routledge, 2000, 32.
55 Ibid., 32.
56 Erich Pauer, 'Preface', *Japan's War Economy*, Erich Pauer, ed. London: Routledge, 1999, xiii.
57 Reiko Niimi, 'The Problem of Food Security', *Japan's Economic Security*, Nobutoshi Akao, ed. New York: St Martin's Press, 1983, 174.
58 Ministry of Agriculture, Forestry and Fisheries, *Nōrinsuisan nenkan, 1951–52* (Annual Report of the Ministry of Agriculture, Forestry and Fisheries, 1951–52). Tokyo: Nihon nōson chōsakai, 1952, 6.
59 Ibid., 8.
60 Fisheries Agency, 'Gyogyō hō shōwa 24 nen' (Basic Fisheries Law 1949), Article 3, found in *Gyogyō kankei hōreishū* (Volume of Fisheries Statutes). Tokyo: Suisan shūhōsha, 1960, 53. It is interesting to note that an article on distant-water fishing was included as early as 1949, given the strong opposition Japan faced from various member nations on the Far Eastern Council (other than the United States) with respect to renewing fishing activities in foreign fishing grounds.

Bibliography

Aki Kōichi. *Nihon no shigen mondai* (Japan's Resource Problem). Tokyo: Kokon Shoin, 1952.

Blakeslee, George. *The Far Eastern Commission: A Study in International Cooperation, 1945–1952*. Far Eastern Series 60. Washington, DC: Department of State, 1953.

Borden, William S. *The Pacific Alliance: United States Foreign Economic Policy and Japanese Trade Recovery, 1947–1955*. Wisconsin: University of Wisconsin Press, 1984.

Buckley, Roger. *Occupation Diplomacy: Britain, the United States and Japan, 1945–1952*. Cambridge: Cambridge University Press, 1982.

Chida, Tomohei and Peter Davies. *The Japanese Shipping and Shipbuilding Industries: A History of their Modern Growth*. London: The Athlone Press, 1990.

Cohen, Theodore. *Remaking Japan: The American Occupation as New Deal*, Herbert Passin, ed. New York: Free Press: 1987.

Esterly, Henry. 'Overseas Fisheries and International Politics in the Occupation of Japan, 1945–52', in *The Occupation of Japan: Economic Policy and Reform*, Lawrence Redford, ed. Norfolk: MacArthur Memorial, 1980.

Fisheries Agency. *Suisanchō 50-nenshi* (Fifty Year History of the Fisheries Agency). Tokyo: Suisanchō 50-nenshi kyōkai, 1998.

George Mulgan, Aurelia. *The Politics of Agriculture in Japan*. London: Routledge, 2000.

Herrington, William. 'In the Realm of Diplomacy and Fish', *Ecology Law Quarterly* vol. 16, no. 1 (1989): 101–18.

Ikeo, Aiko. 'Economists and Economic Policies', in *Japanese Economics and Economists Since 1945*, Aiko Ikeo, ed. London: Routledge, 2000.

Kawakami Kenzō. *Sengō no kokusai gyogyō seido* (The Post-war International Fisheries Regime). Tokyo: Dai Nihon Suisankai, 1972.

Ministry of Foreign Affairs (MOFA). *Kita taiheiyō no kōkaigyogyō ni kansuru kokusai jōyaku kankei iken* (Opinions Regarding International Agreements on National Fisheries in the North Pacific), Microfilm Series B'0039, MOFA Archives, n.d.

Niimi, Reiko. 'The Problem of Food Security', in *Japan's Economic Security*, Nobutoshi Akao, ed. New York: St Martin's Press, 1983.

Okita, Saburo. *Postwar Reconstruction of the Japanese Economy*. Tokyo: University of Tokyo Press, 1992.

Pauer, Erich. 'A New Order for Japanese Society: Planned Economy, Neighbourhood Associations and Food Distribution in Japanese Cities in the Second World War', in *Japan's War Economy*, Erich Pauer, ed. London: Routledge, 1999.

Schaller, Michael. *The American Occupation of Japan: The Origins of the Cold War in Asia*. Oxford: Oxford University Press, 1985.

Scheiber, Harry. 'Pacific Ocean Resources, Science, and Law of the Sea: Wilbert M. Chapman and the Pacific Fisheries, 1945–70', *Ecology Law Quarterly* vol. 13, no. 3 (1986): 381–534.

——Inter-Allied Conflicts and International Law, 1945–53: The Occupation Command's Revival of Japanese Whaling and Marine Fisheries. Taipei: Academia Sinica, 2001.

Selak, Jr. Charles B. 'Recent Developments in High Seas Fisheries Jurisdiction Under the Presidential Proclamation of 1945', *American Journal of International Law* vol. 44 (October 1950): 670–81.

Zacher, Mark W. and Brent Sutton. *Governing Global Networks: International Regimes for Transportation and Communications*. Cambridge: Cambridge University Press, 1996.

3 Negotiating a regional fisheries system in the North Pacific

Food security planning mandated renewed support for the notion of 'open' fisheries management for Japan and adherence to pre-war concepts of *mare liberum* despite greater calls for restrictive maritime zones. As will be seen in this chapter, Japan, the United States and Canada concluded Japan's first post-war fisheries treaty, the International North Pacific Fisheries Convention (INPFC), and it purposefully reaffirmed the principle of freedom of the seas upon which other subsequent bilateral fisheries agreements in the North Pacific were negotiated. In effect, it helped establish an open oceans system in the post-Occupation period that facilitated the rehabilitation and eventual expansion of Japanese distant-water fisheries. Subsequent negotiations with the Soviet Union, the People's Republic of China (PRC) and South Korea were beset by many problems, such as the absence of formal governmental ties, wartime resentment or the absence of a peace treaty, but each of these barriers was surmounted in turn. Frequently, Japanese industry representatives acted as intermediaries and were able to offer various economic cooperation agreements in conjunction with much coveted fisheries treaties.

Tripartite negotiations for a North Pacific fisheries treaty[1]

Since traditional fishing grounds near Korea, China and the Soviet Union remained closed due to wartime resentment, security considerations and the fears of fish stock over-exploitation, the Supreme Command of the Allied Powers (SCAP) authorities recognized that they would have to open up fishing grounds to the east of Japan, in the North Pacific, to allow Japanese fishers to expand their catch to meet domestic needs. By 1949, Washington and Tokyo felt that an extension of Japan's fishing areas could best be accomplished with a precedent-setting multilateral agreement that set out rights and rules governing access in the North Pacific.

American fishing interests as well as Australian, Korean, Indonesian and Chinese representatives pressured SCAP and the State Department to enshrine restrictions on Japanese fishing in the Peace Treaty.[2] Somewhat alarmed by the extent of these requests, Ambassador Dulles even commented that the treaty was becoming an 'international fisheries convention'.[3] He opposed such attempts largely to sidestep any singular opposition this might bring to the treaty as a whole and to avoid a punitive settlement with respect to Japan's international fisheries. In order to conclude the treaty as soon as possible, 'this pressure had been resisted in every instance'.[4]

To ensure that restrictions were not imposed upon Japan in the Peace Treaty, Prime Minister Yoshida made special efforts during discussions leading up to the final draft to reassure Ambassador Dulles that the interests of American salmon fishers would be respected by Japanese fishermen in the post-treaty era. In a personal letter to Dulles dated 7 February 1951, Yoshida pledged that the 'Japanese government will, as a voluntary act, implying no waiver of their international rights, prohibit their resident nationals and vessels from carrying on fishing operations in *presently conserved fisheries* ... and in which Japanese nationals or vessels were not in the year 1940 conducting operations'.[5] Since only salmon, halibut and herring fisheries in Alaska and the East Pacific fitted the definition of 'conserved', this letter served as a declaration of Japan's voluntary abstention from these American and Canadian fisheries.[6] Nonetheless, aware that such an abstention might serve as an international precedent affecting Japanese interests in other fishing grounds, Yoshida was careful to qualify this commitment by saying that it would not imply a 'waiver of international rights'. Japan was firmly committed to defending freedom of the seas in its future negotiations.

The final draft of the Peace Treaty included a provision on fisheries, but it set no limitations or boundaries upon Japanese fishers as hoped by many Allies and American fishing interests. Instead, Article 9 of the Peace Treaty simply stated that 'Japan will enter promptly into negotiations with the Allied Powers so desiring for the conclusion of bilateral and multilateral agreements providing for the regulation or limitation of fishing and the conservation and development of fisheries on the high seas'.[7] Thus, Japan was required to negotiate bilateral or multilateral fishing treaties before being allowed to fish near the coastal waters of respective countries.

Concurrent with negotiations for the Peace Treaty, a separate agreement was under discussion in 1952 as a first step in Japan's general effort to conclude fisheries treaties with surrounding countries. Japan and the United States decided to negotiate the first fisheries treaty in an effort to establish a favourable precedent for ensuing negotiations as well as swiftly to open up an abundant fishing area for Japanese fishers.[8] Washington supported the conclusion of a treaty in the North-East Pacific, with the inclusion of Canada, since an agreement for the Western Pacific and the Japan and China Seas was still nearly impossible.

Tripartite negotiations were held in Tokyo in November and December 1951. SCAP conveyed temporary sovereignty to Japanese authorities for the duration of the negotiations since the Peace Treaty had not yet been completed and full sovereignty had yet to be restored.[9]

The negotiation policy of the Japanese Ministry of Foreign Affairs (MOFA) was to defend the principle of freedom of the high seas by avoiding any restrictions on fishing activities. On 6 October 1951, MOFA circulated a draft policy entitled 'Policy with Respect to Fisheries Treaties', which centred its arguments on the precedent issue: 'The utmost must be done to avoid any agreement which limits or prohibits fishing on the high seas.'[10] It counselled negotiators to avoid accepting any treaty language that might be used to Japan's disadvantage in later negotiations with other nations, because such demands would 'inevitably follow'.[11]

Moreover, Japan wished to secure American and Canadian support of freedom of the seas through a mutual declaration that defended this freedom as a universal right. In the Japanese draft of the convention, Article 2 stated that 'no country concerned under this convention is to be subject to discriminatory exclusion from the exploitation of any high seas fishing resource ... '[12] In other words, the Japanese argued that no restrictions should be applied to any single country in compliance with the standards of the earlier oceans regime, while conservation measures should be applied equally to all signatories. The Japanese negotiators hoped to avoid the adoption of any exclusionary principles for fear they might serve as adverse precedents in later negotiations with the USSR, the PRC and South Korea.

MOFA also emphasized the particular importance of fisheries to the economic revitalization of the post-war economy. The Fisheries Agency had circulated a policy paper in October 1951 that argued for the support of freedom of the seas in the context that undermining this principle would also undermine Japanese distant-water fisheries, which would have devastating effects upon Japan's attempts at self-sustaining economic development.[13] This tactic was persuasive since Washington had already committed itself to the reconstruction of Japan through SCAP while the State Department wished to divest itself from financially burdensome aid expenses through the promotion of a self-supporting domestic economy. Moreover, Japan and the United States were concurrently negotiating arrangements for a mutual security treaty whereby the 'economic stability and development of Japan shall be its prerequisite ... '[14] The reasoning behind such a strategy was that fisheries were an important component of Japan's economic development and, consequently, of American security interests.

Canadian and US fishing industries, on the other hand, sought to exclude Japan from any fishing in the North-East Pacific once they realized that the Truman Proclamation would not be enforced. As resolved in the November 1950 meeting of the Pacific Fisheries Conference, which was made up of delegates from American and Canadian industrial fishing companies, West Coast fishers encouraged their respective governments to negotiate with the objective of ensuring that 'Japanese fishermen will stay out of the fisheries of the Northeast Pacific Ocean which have been developed and husbanded by the United States and other countries of North America ... '[15] Fishing industry advocates cited Japan's earlier 'invasion' of the Bristol Bay salmon fishery in 1936–37 as well as potential threats to peaceful relations with Japan as substantial reasons to impose an eastern limit on Japanese distant-water fishing.[16]

The US government, however, concurred with MOFA's position with respect to the defence of freedom of the seas and the need for economic revival through fisheries expansion. Much more concerned with communist victory in China, Soviet expansionary policy in East Asia and the Korean War than with fisheries in the Bering Sea, the State Department was attentive to security needs in Asia. As stated in a US Central Intelligence Agency special estimate conducted for the State Department in 1951, Japan would play a critical role in establishing an 'East-West balance of power in the Far-East' and it would be important to secure markets and natural resources, such as fish, in non-communist areas.[17] The fall of Nationalist China in 1949 also raised urgent problems regarding the replacement of the very important and proximate mainland economy with one large enough to support the Japanese market, while remaining friendly to American interests.

Part of the answer to this dilemma was to open American markets to Japanese exports to provide a source of dollars for the capital-poor Japanese economy, as in the case of tuna fisheries.[18] This policy was adopted in part to prevent Japan from turning to the market of communist China in search of export earnings. Another partial solution was to allow the Japanese into the North Pacific to secure access to another source of natural resources. The Tripartite negotiations were conducted with the understanding that the United States and Canada would allow Japan to fish in waters proximate to American coastal waters while making efforts to prevent Japan from entering selected, valuable North American fisheries, namely salmon, halibut and herring. Thus Japan would be able to develop an export market and provide for domestic consumption needs without having a directly adverse affect on valuable North American fisheries.

While the contemporary oceans regime did not allow the exclusion of Japan from such fisheries, the North American delegations pressured the Japanese to adopt some *voluntary* measures to reserve these stocks for Canadian and American exploitation. MOFA and the Ministry of Agriculture, Forestry and Fisheries (MAFF) agreed to a compromise whereby an abstention principle would be applied to Japanese fishers alone.[19] Although the adoption of such a restriction may have seemed severe in terms of Japan's prewar experience, the agreement nonetheless opened up a vast fishing area beyond the diminutive MacArthur zone for future use.

The resulting agreement, known as the International North Pacific Fisheries Convention, was finally signed on 9 May 1952 and came into effect on 12 June 1953.

The INPFC contained three important provisions. First of all, support for freedom of the high seas was included in the preamble in accordance with Japan's wishes.[20] Second, the abstention principle was adopted in full, but included in the appendix of the treaty. While not readily apparent, this section served as the operating basis for the new treaty, whereby the Japanese in effect agreed not to fish herring, salmon and halibut east of an Abstention Line demarcated at 175 degrees west longitude. Finally, the agreement established a supervisory commission to provide an annual trilateral forum to coordinate dialogue and to encourage biological studies on designated stocks.

Unlike several other extant fisheries organizations, such as the International Whaling Commission, the INPFC was relatively decentralized. It did not formulate quotas, could not enforce decisions upon its members, and did not have a formal dispute-resolution process.[21] While the INPFC was one of the first fisheries treaties to supervise scientific studies and regularly employed conservationist language at annual meetings, it was not a conservation-minded organization. Allocating salmon, herring and halibut was its effective function, not the management of environmental sustainability. Instead, conservation considerations were the voluntary responsibility of each national government.

The most important role of the INPFC was to re-open the North Pacific to Japanese fishing after a period of restriction imposed by the MacArthur Line. The INPFC endorsed freedom of the seas as the governing principle for fisheries in the North Pacific, which was the international system best suited for Japan to achieve its goal of food security. The INPFC also set an important precedent for subsequent bilateral negotiations that were conducted with Japan's neighbours in the following two decades, most importantly the Soviet Union, the PRC and South Korea. The remainder of this chapter will examine the background to these negotiations and the resulting agreements.

Bilateral fisheries treaty with the USSR

The Pacific War also entailed a realignment of fishery relations between Japan and the Soviet Union. Both countries had concluded two previous fisheries agreements in 1925 and 1928, but these were nullified as a consequence of the war and the new boundaries it produced.[22] The Soviet Union had occupied the Japanese Kurile islands including Habomai and Shikotan at the conclusion of the Pacific War, and Moscow's insistence on a 3-mile territorial limit around these newly acquired territories deprived Japan of several productive fishing grounds that were traditionally relied upon.[23]

With the conclusion of the INPFC and the lifting of the MacArthur Line restrictions that opened the North Pacific to Japanese fishers in 1952, Japan also wished to renew fishing operations near the Soviet Union's eastern borders, in a body of water the Japanese refer to as the 'Northern Seas'. Tokyo thus hoped to start talks aimed at concluding a fisheries treaty with Moscow in the shortest term possible. Since the USSR did not participate in the San Francisco Peace Treaty process, however, Japan had to negotiate concurrently a separate normalization agreement and a fishing treaty with Moscow.

Japan first approached the Soviet Union with a dispatch of leading company representatives who met with Soviet officials in Moscow as early as

January 1952 in an attempt to secure permission to fish in the Northern Seas.[24] Officials from both countries were unable to agree on fishing boundaries near the disputed islands in the north, however, and the exploratory discussions ground to an early halt.[25] This stumbling block foreshadowed the territorial dispute that has plagued Japanese-Russian relations to this day.

Despite the absence of a fishing agreement with the USSR, the Japanese government authorized a succession of mothership fleets to explore the potential of salmon fishing in the Okhotsk Sea and the western Bering Sea between 1952 and 1956.[26] These fishing ventures near Soviet coastal waters proved to be controversial in Moscow, perhaps forcing the hand of Russian authorities to negotiate some kind of agreement in order to reassert their authority over the eastern coastline and to set controls on these fishing operations. Japan had demonstrated that fishing could take place even in the absence of an agreement.

Japan continued efforts to restart negotiations with the Soviet Union for a fisheries agreement while fishing companies played a leading role in establishing bilateral contacts. The Japanese fishing industry worked through the 'National Council of Associations for the Normalization of Relations with the Soviet Union and China' in an attempt to bring these governments together.[27] In August 1953, Onishi Kensaku, President of Hokuyō Suisan Company, travelled to the Soviet Union and reported back to Tokyo that trade relations with the USSR could be developed even before formal normalization.[28] This reflected, in part, the desire of fishing companies to combine trade and fisheries questions and to separate territorial issues from fisheries issues. For its part, authorities in the Soviet Union sought trade linkages with Japan and were prepared to offer a fisheries agreement in return for Japanese economic cooperation.

When the Hatoyama government took office in December 1954, it was favourably inclined toward normalizing relations with the Soviet Union and soon initiated negotiations for normalization and fisheries. Talks were held on 3 June 1955 in London between the Japanese ambassador to the United Kingdom and the Soviet foreign minister, where both negotiators agreed at the outset upon two provisions: a fisheries treaty would be negotiated after the conclusion of a peace treaty, and it would endorse the creation of a bilateral organization responsible for the rational utilization of fisheries.[29] After nine months of discussion, however, the negotiations finally stalled over the question of territorial boundaries in the Russian Far East, and the hope of finding an early solution to the peace treaty problem was abandoned.

On 21 March 1956, the day after normalization negotiations were suspended, the Soviet Union announced its intention to establish a 'conservation zone' to protect salmon and trout stocks contiguous with its border.[30] The Soviets drew a demarcation line, known as the Bulganin Line, encompassing the Okhotsk Sea and waters adjacent to the Kamchatka Peninsula that aimed to deny entry to Japanese fishing trawlers.[31] The Soviets were becoming increasingly worried about Japanese salmon operations off their eastern coast, which they declared were using 'reckless' practices.[32] The unilateral nature of the Soviet declaration and its timing reinforced the importance to the Japanese of the advantages of resuming normal relations with the Soviets.

It may be noted that while the official reason for this proclamation was over environmental concerns, the Okhotsk Sea was also home to the Russian Pacific Naval fleet, which was based in Vladivostok, and the establishment of such a line served to prevent access to vessels that might be in the position to gather intelligence.[33] Given that the Bulganin Line did not protect much of the overall migration distance of Russian salmon, the security element may have been the operational one. This was certainly true of the 'no-sail' and 'no-fly' zone that the Soviets declared for Peter the Great Bay in 1957.[34]

The declaration of the Bulganin Line was a serious threat to Japanese fishing interests and the industry soon reapplied pressure for a swift end to the negotiation deadlock. The Japanese side sought to distance itself from the territorial issue in an effort to reach even a partial agreement on fisheries. New negotiations, which were limited to fisheries issues alone, were restarted in Moscow on 29 April 1956.[35] The chief negotiator for the Japanese side, Kōno Ichirō, the minister for agriculture, forests and fisheries, was a boon to fishing interests due to his special connection with the industry, a one-time fisheries specialist for the *Asahi Shinbun* and a recipient of political donations from the industry.[36] He used his influence to bypass possible opposition from his own party and MOFA to negotiate directly with his Soviet counterparts, reaching an agreement that endorsed the reopening of peace negotiations by July that same year in exchange for a fisheries treaty.[37] An agreement for the normalization of relations, in the absence of a peace treaty, was finally announced on 12 December 1956 – the same day the new fishing treaty came into effect.[38]

In this way, pressure from the Japanese fishing industry for the conclusion of a fishing treaty created the impetus for a breakthrough in the peace negotiations.[39] The fishing agreement overall may be seen as having played an important role in normalizing relations between the two countries in the absence of a peace treaty and this system remained in effect until it was replaced in 1978.[40] As a result of industry efforts to initiate normalization talks prior to the conclusion of the agreement, these negotiations may also be seen as an example where domestic fishing interests played a role in aiding and formulating Japanese foreign policy.

The final version of the fishing treaty permitted the Japanese to fish for salmon, crab and herring in specially demarcated zones outside the Bulganin Line under regulations set by the Japanese government.[41] Since the Russians did not fish near the Japanese coastline and thus could not be offered reciprocal access to Japanese fishing grounds, the Soviets sold fishing rights directly to the Japanese government. A committee composed of three members from each country was established to negotiate the allowable catch levels for Japanese salmon fishers on a yearly basis. It is significant to note that the treaty did not contain an 'abstention' clause as feared by the Japanese negotiators.[42] The committee's talks focused on the extension of areas where fishing for salmon and crab were prohibited, the maximum annual catches, the duration of the fishing season, and the length, spacing and mesh size of driftnets.[43]

The treaty also called for allocation of fish stocks to be determined on the basis of scientific management. However, according to a model of inputs and outputs of yearly negotiations between 1956 and 1976 developed by Takashi Inoguchi and Nobuharu Miyatake, one of the operative components that helped set allowable catches was the price index for the Tokyo-based Tsukiji fish market.[44] Thus it would appear that while the treaty specified scientific research as the determiner of catch levels, the more important element was the Japanese domestic market. Another factor prompting more generous levels of fisheries allocation was the ever-present threat that small allocations would simply lead to greater unlicensed fishing on behalf of the Japanese.

The most valuable catches in the North-West Pacific included Alaskan pollack, salmon, crab, squid and herring, although the area was abundant in

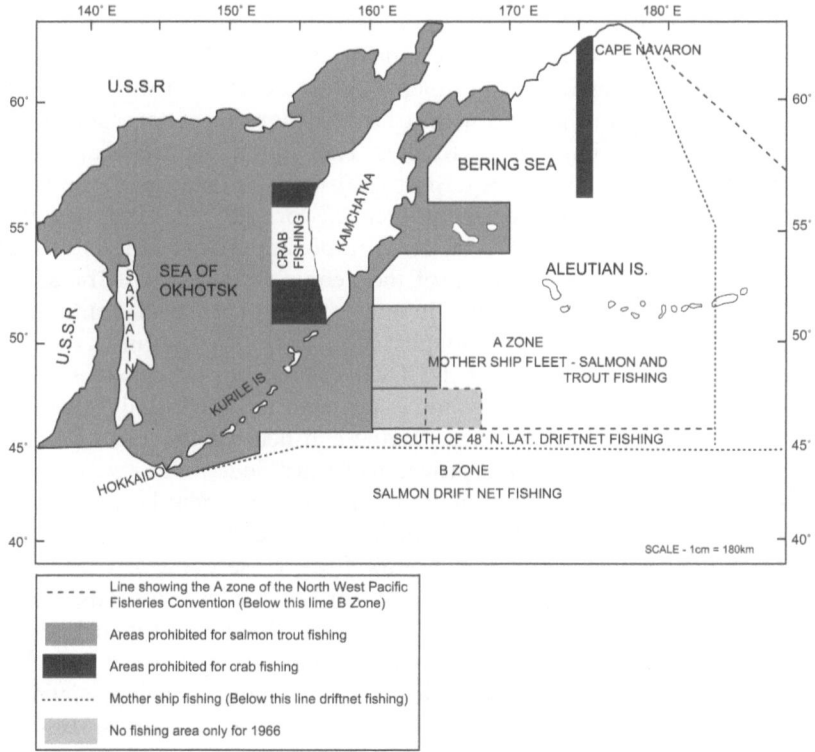

Figure 3.1 Map of areas prohibited for fishing as agreed to by Japan and the Soviet Union

Source: Savitri Vishwanathan, *Normalization of Japanese-Soviet Relations, 1945–1970.* Tallahassee, FL: The Diplomatic Press, 1973, 54

many other species as well. Japan's total salmon catches in 1955–61 were larger than those of the United States, and in some years larger than the combined catches of both Canada and the United States.[45] Thus the convention opened up a vast new source of salmon and other commercially valued species for the Japanese market despite the application of some restrictions imposed by the Bulganin Line.

Again, Japan was able largely to surmount the threat of exclusionary fisheries restrictions while obtaining the Soviet Union's endorsement of the principle of freedom of the seas. This was accomplished through a combination of private industry initiatives and the political will of bureaucrats and politicians who favoured the normalization of relations with the Soviet Union. This style of diplomacy was characterized by a non-ideological, practical approach to negotiating differences in the economic sphere despite the political tensions that existed between these two Cold War adversaries.

Bilateral fishing agreements with the People's Republic of China

With the lifting of the MacArthur Line restrictions in 1952, Japan was eager to re-enter the productive fishing grounds of the East China and Yellow Seas. In accordance with Japan's commitments under Article 9 of the Peace Treaty, Tokyo wished to pave the road with a fisheries agreement to ensure the safe entry of its trawlers into these grounds. The diplomatic landscape in East Asia, however, had undergone a seismic shift with the creation of the PRC on 1 October 1949, obstructing many avenues for dialogue and negotiation between Tokyo and Beijing.[46] The main impediment was Japan's refusal to acknowledge officially the creation of the People's Republic of China, which held de facto control over the Chinese mainland, instead continuing to recognize Chiang Kai Shek's Nationalist government in Taiwan as the official government of China. This created a legal vacuum for the Japanese government when trying to solve diplomatic issues with the PRC, and Tokyo feared that this would create trouble for Japanese fishers in the East China and Yellow Seas.[47] Indeed, this soon proved to be justified.

Despite the absence of an international fishing agreement between the PRC and Japan in 1952, Japanese trawlers had been conducting limited fishing operations in the East China and Yellow Seas, close enough to the coastal waters of Mainland China to raise the concerns of Chinese officials. At times, these fishing fleets came into conflict with the Chinese coast guard, who disputed the legality of these operations and cited various reasons – including fisheries violations, interference with domestic Chinese fishing, spying and MacArthur Line infractions – to justify impounding numerous vessels and scores of fishermen.[48] Between 1950 and 1954, the PRC seized a total of 158 Japanese fishing vessels operating in the East China and Yellow Seas, and detained a total of 1,909 fishermen.[49] The extent of the seizures was enough to raise protests by Japanese fishing companies, who appealed to the Japanese government to find a way to repatriate vessels and crews, and to secure uninhibited fisheries access in these fishing grounds. MOFA, however, refused to initiate official contact with a regime it regarded as illegitimate and illegal.[50]

Although it would appear as though tense political relations between Japan and communist China at the official level might serve as a stumbling block to fisheries negotiations, the Japanese fishing industry managed to mobilize domestic politicians to initiate informal talks. The first overtures to establish direct negotiations with the PRC were made in 1952 by the PRC-Japanese Fishing Enterprisers' Association who were hoping to capitalize on the good relations derived from the recent conclusion of the private trade agreement with China in the same year.[51] It was not until October 1954, however, that the PRC gave an affirmative response. During a visit to Beijing by a group of Japanese Diet members, including Taguchi Chōjirō and Matsuura Seiichi, Chinese Premier Chou En Lai offered to establish negotiations between non-governmental members of both nations to negotiate a fisheries agreement and discuss the repatriation of seized vessels.[52] In a display of its goodwill toward the negotiations, Beijing stopped the seizure of Japanese fishing vessels, and all crew members and thirty fishing boats were repatriated to Japan by the end of 1954.[53]

The Japanese delegation, the PRC-Japanese Fishery Council, was established on 13 November 1954 and began talks with their Chinese counterparts, the PRC Fishery Association, on 13 January 1955.[54] The ensuing negotiations involved heated exchanges over three months and despite a number of occasions when they appeared likely to collapse, the 'Agreement Concerning Fishery in the Yellow and East China Seas between the PRC-Japanese Fishery Council of Japan and the PRC Fishery Association' was finally concluded on 15 April 1955.[55]

The Japanese delegates had entered the negotiations with two principal objectives in mind. The first objective was to have the PRC recognize the principle of freedom of the high seas as the operating basis of the agreement, much as the Japanese government had done in previous negotiations with the Soviet Union and the United States/Canada. The second objective was to secure Chinese commitment to 'safe fisheries', which for the Japanese meant that fishing boats should be free to catch fish without being subject to seizure.[56] Again, Japanese delegates were careful not to support any terminology in the final agreement that would deprive Japanese fishers exclusively from oceanic fishing grounds.

The PRC negotiators, on the other hand, sought to create several prohibited areas near their coasts in accordance with previously promulgated domestic law. Perhaps owing to the pressures of Japanese fishing fleets near their coastal waters, the Chinese communists laid claim to territorial waters and contiguous fishery grounds in December 1950 – only a year after coming into power – through the 'Regulation on Trawl Fishery Operation in East China'.[57] The boundaries of this claim, however, were not explicitly asserted until the negotiations. The Chinese delegates argued for the creation of three military zones that would be off limits to Japanese fishers and one conservation zone demarcated by what has been termed the Mao Tse-tung Line, in the style of the MacArthur, Bulganin and Rhee Lines that preceded it.[58]

The final agreement mirrored in many ways the earlier tripartite agreement signed by the United States, Canada and Japan. The main text contained the requisite provision that affirmed both parties' commitment to freedom of the seas, and laid the foundation for six fishing zones off the Chinese coasts that delineated the sizes of fleets and length of seasons.[59] Meanwhile, Japan agreed to exclusive measures that established prohibited zones for security and conservations reasons, but this was done informally in an exchange of letters after the signing of the original agreement.[60] Japan, in effect, resigned itself to accepting a smaller abstention zone in which they would prohibit the operation of their own fishers in exchange for access to uninhibited fishing rights in much larger fishing grounds. This was also agreed to in a manner that would avoid setting an international precedent for exclusive fishing zones.

Although this agreement seemed to clear the path for improved relations between the PRC and Japan, fisheries relations between the PRC and Japan conformed to the state of their political relations. The fishery agreement was meant to be provisional and Article IX of the agreement encouraged both governments to open negotiations to conclude a more binding fishery agreement.[61] Government-level negotiations, however, did not occur until 1975. Although the 1955 non-governmental agreement was extended for two more years, a chill in PRC-Japan political relations in 1957–58 led the Chinese Council to notify its Japanese counterpart that it wished to terminate the fisheries agreement in June 1958.[62] Prime Minister Kishi's ascension to power in 1957 shifted Japanese policy in a strongly pro-Taiwanese and anti-communist direction that raised the ire of Beijing. The PRC finally used an incident in May 1958 when a Japanese right-wing nationalist burned a PRC flag in Nagasaki as a pretext to deprive Japan of the fishing agreement it desired.[63] It was also during this time that the PRC began to seize Japanese fishing vessels again.[64]

A five-year period followed with no new agreement. Owing in part to the new Ikeda administration's emphasis on *seikei bunri*[65] and a favourable orientation toward Beijing, the PRC agreed to new fisheries negotiations with Japan in November 1963. An agreement was concluded that same month and contained basically the same provisions as its 1955 predecessor. This agreement remained in force, with some modifications to common fishing zones in 1965, until formal diplomatic recognition was offered to the PRC in 1972.[66] Hereafter, Japan and the PRC concluded a formal governmental treaty that took effect in December 1975.

Since no formal relations existed between the PRC and Japan prior to 1972, all three fishing agreements concluded between the countries were negotiated by non-governmental organizations and all were private in nature. The agreements did, however, have a de facto influence on the policies adopted and enforced by each respective government and were thus an important contribution to help normalize fisheries relations between Tokyo and Beijing.

Fisheries relations, however, proved subsidiary to the changing climate of political relations between the governments of Tokyo and Beijing. Whereas fishing industry officials and sympathetic politicians effectively employed a negotiation strategy to bypass political impediments in the realm of fisheries, political relations at the state level took several decades to resolve.

Bilateral fisheries treaty with South Korea

The fisheries near the Korean Peninsula were of particular importance to Japan in the pre-war years. In the 1930s, Korean and other colonial fisheries supplied close to one-third of Japan's total production. Thus the re-opening of these fishing grounds was seen as key to the prosperity of Japan's post-war fisheries.

The MacArthur Line prevented Japanese trawlers from returning to Korean coastal waters during the Occupation, but as SCAP's term drew to a close, Korean officials worried that Japanese trawlers would return to their fishing grounds in force. While the Peace Treaty was being finalized, Korean officials approached the Japanese government to propose early negotiations for a fisheries treaty in accordance with the proposed Article 9 in October 1952.[67] It was hoped that an early settlement might limit the resumed Japanese fishing effort near Korean waters and set in place the necessary controls before the MacArthur Line was lifted. Despite having agreed to early negotiations with the United States and Canada at this time, Japan ignored Korea's request for the same.[68]

In the view of the Korean government, fears of a revived Japanese presence in Korean fishing grounds were justified. The Korean government charged that Japanese fleets had pillaged its fishing grounds by indiscriminate over-fishing throughout the 1930s to the extent that sea resources and feeding grounds had become depleted. Korean officials also regularly raised the spectre of Japanese overfishing in Korean waters as evidence of the need to place fishery restrictions on Japan in the Peace Treaty.[69] The strongest evidence of Japan's determination to return to their former fishing grounds was the presence of many Japanese trawlers in Korean waters in contravention of the limits imposed by the MacArthur Line. By the end of 1952, the Korean Navy had detained 108 Japanese vessels in violation of SCAP limits.[70]

Perhaps owing to these fears and the desire to protect Korean fishers from renewed competition from Japanese fishing fleets, President Rhee unilaterally issued the 'Presidential Proclamation of Sovereignty over the Adjacent Sea' on 18 January 1952, otherwise known as the 'Rhee Line'.[71] In the absence of a fishing treaty with Japan, this proclamation laid claim to waters contiguous to the Korean coast and prohibited foreign fishing fleets from entry.[72] The Korean government justified its proclamation by citing the Truman Proclamation of 1945 as a precedent.[73] Japan, followed by the United States, Britain and Taiwan, soon objected to the declaration as being contrary to international law and the commonly held principle of *mare liberum*.[74] The Japanese government must have been particularly troubled by the fact that this was the first such exclusive fisheries claim by one of Japan's close neighbours and as such could have served as a precedent for future proclamations by other states.

Congruent with Korean efforts during the Occupation to place permanent restrictions on Japanese fisheries, the Rhee Line sought to protect coastal fisheries for the exclusive use of Korean fishers.[75] The declaration also served an important political purpose. The Koreans had reason to fear that Japan might try completely to avoid negotiating a fisheries treaty since this would permit Japanese fishers unrestricted, unregulated access to the Japan Sea and areas contiguous to Korea.[76] In this case, the unilateral declaration would encourage the Japanese to join the negotiating table and would also serve as a powerful bargaining tool during subsequent talks.

The Japanese and Koreans hotly contested the validity of the Rhee Line in the following months. Hostilities between the Japanese and Koreans over fisheries flared to the extent that in September 1952 General Mark Clark, Supreme Commander of United Nations forces in Korea and also Commander of US Forces in East Asia, created a 'defence zone' surrounding Korea that prevented any vessels from sailing in or out of Korean coastal waters.[77] While this zone was ostensibly created to prevent communist infiltration into the areas where Allied forces staged operations, he privately declared it a 'fisheries zone' designed to keep the Japanese and Koreans from coming to blows over fishing rights.[78]

No fewer than six rounds of fisheries negotiations were attempted by Korea and Japan between 1952 and 1964, forming part of an overall round of normalization talks that also related to, among other things, Korean residents in Japan and Japanese property claims in Korea. According to Article 21 of the Peace Treaty, Korea was entitled to the benefit of Article 9 requiring Japan to negotiate a fisheries treaty with surrounding countries.[79] Fisheries talks thus formed an integral part of the larger negotiations relating to normalizing relations between the countries.

The first talks held on 20 February 1952 (within a single month of the Rhee Line proclamation) laid the seeds of future disagreement. In a policy line that would be repeated throughout subsequent negotiations until 1964, the Korean delegation sought Japan's recognition of the Rhee Line and its claims to various territorial waters.[80] Japan, on the other hand, refused to accede to President Rhee's coastal water claims, and instead argued that both countries should uphold the principle of freedom of the seas while fisheries in the Japan Sea (which the Koreans called the Korean Sea) should be jointly managed.[81] As with negotiations with the PRC, it also argued for the support of 'safe' fisheries, meaning that fishing vessels would be able to operate without fear of seizure.[82] Each time, however, the negotiations faltered on the territorial claims issue.[83] It should be noted that these negotiations took place in the context of a generally hostile relationship between the two countries,

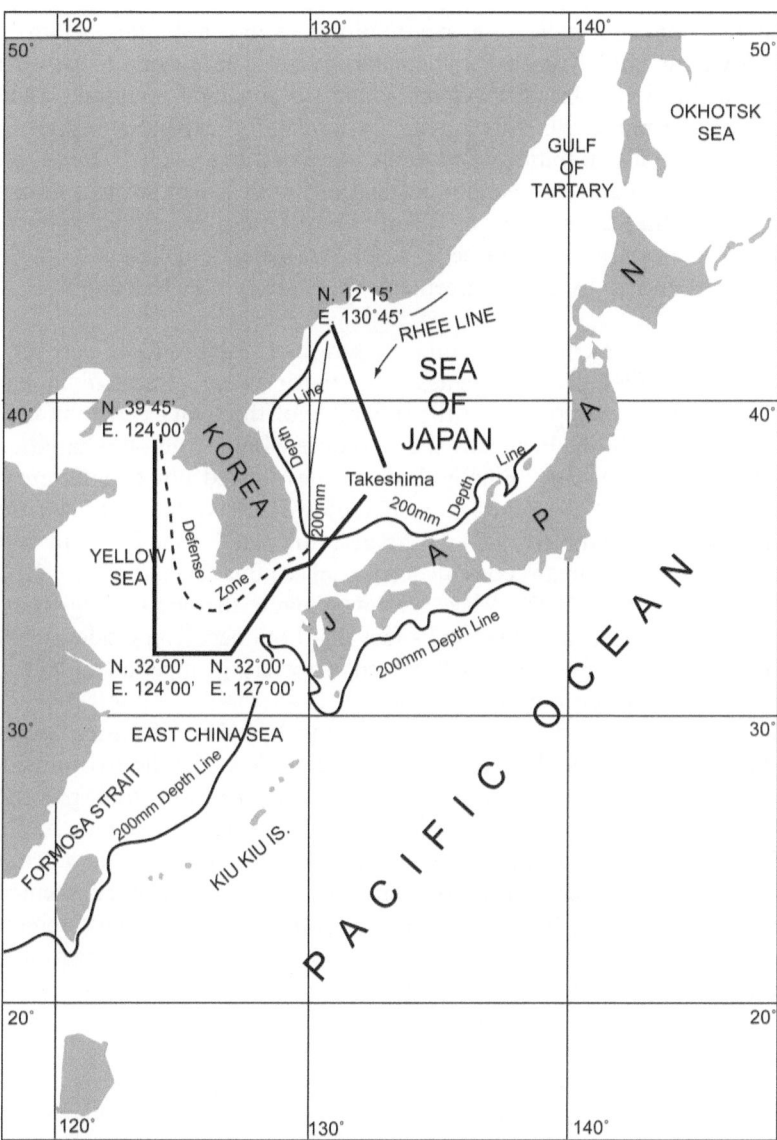

Figure 3.2 Map of Korea, Japan and the Peace Line
Source: Shinji Taguchi, 'Japanese-Korean Fishery Dispute', *Contemporary Japan* vol. 22 (1953): 398

typified by President Rhee's unflagging enmity toward Japan.[84] This made the normalization and fisheries talks particularly problematic and unusually bitter.

Fisheries and normalization negotiations would not start again with renewed vigour until Korean President Park Chung Hee came to power in a *coup d'état* in 1961. The Park regime sought to conclude economic, fisheries and other normalization agreements in return for Japanese support and capital to fuel his ambitious economic development plans.[85] Once South Korea decided to enhance economic relations with Japan, it was willing to subordinate fisheries interests in favour of reaching an overall agreement. Negotiations hereafter took place in a spirit of genuine willingness to seek a compromise and a fisheries agreement was finally concluded on 22 June 1965.[86]

The resulting agreement contained a series of compromises between the two countries. First of all, Japanese negotiators were successful in having freedom of the seas supported in the preamble of the treaty.[87] A joint control zone was also established whereby the resources within would be shared on an equal (i.e. fifty-fifty) basis. Finally, both parties agreed not to interfere with the fishing vessels of the other, even if they were in violation of the treaty, which served to guarantee Japan's request for 'safe' fisheries.[88] In return, Japan recognized Korean coastal territorial claims extending to 12 nm, although no mention of the Rhee Line was made, thereby leaving its status ambiguous.[89] It was largely due to this fact that both negotiating sides were able effectively to surmount what had proved a towering impediment and thus settle an unusually acrimonious dispute.

The fisheries treaty between Japan and the Republic of Korea seems to have been made possible largely by the new willingness of the Korean side to offer concessions. Broader political-economic considerations took precedence over the rigid adherence to the fishery restrictions that were a source of continuing grievance with Japan. Once again Tokyo was able to take advantage of the desire of their negotiating partner to achieve an economic agreement in order to conclude the fisheries treaty it desired. The fisheries agreement proved durable over the years, with neither side proposing an abrogation even when afforded the opportunity in 1970.[90]

Conclusion

Several observations may be made about the bilateral phase in Japan's foreign fisheries relations in the North Pacific.

First of all, the United States served as the setter of precedent in oceans regime making. Coastal claims, including the Rhee Line and the Bulganin Line were often justified in terms of the precedent set by the Truman Proclamation. Although the United States seemed to have reversed course with this declaration, it did stand as an alternative to the *mare liberum* system. Moreover, all fisheries agreements declared their support for freedom of the seas much as the INPFC did, where it too played the role of establishing international precedent in the North Pacific.

Moreover Japan was effective in using a non-ideological approach with its dealings with the Soviet Union and China, despite being antagonists during the Cold War. This permitted Japan to bypass contentious political issues such as territorial claims or formal recognition of national status, and to focus on the fisheries issues at hand. Thus Japan was able to secure access to fisheries resources that might have otherwise been more vehemently protected by these countries.

Another point that should be noted is that the private sector featured prominently in negotiations where formal contacts and intergovernmental diplomacy was problematic or non-existent. This includes negotiations with the PRC, the Soviet Union and South Korea whereby private sector linkages either initiated formal talks or negotiated an agreement in place of normalization. These stand out as flexible responses where political intransigence threatened to halt negotiations.

Another example of flexible response on behalf of the Japanese is that fisheries were rarely negotiated on their own but instead were often linked to other issues such as trade normalization or economic agreements. This permitted Japan larger concessions than they might have otherwise been able to attain, given that Japan could not offer a reciprocal agreement on fisheries where other countries did not seek to fish in Japanese waters. This allowed for horse-trading on connected issues that gave Japan a distinct advantage whereas treatment of fisheries as an issue on its own would have left Japan at a disadvantage. Fisheries ranked highly in the order of preferences for agreement relative to other issues. It should also be mentioned that Japan's neighbours frequently used the confiscation of Japanese vessels and capture of crew as a negotiating ploy in fisheries negotiations. The Japanese were always aware that safe fishing operations required the guarantees from coastal countries to allow its nationals to go unmolested in international waters.

While some of the restrictions may appear harsh in terms of Japan's pre-war experience of completely unrestricted fishing, the agreements were generally liberal in their application and rarely discriminatory over species coverage (with the exception of salmon). They opened up the world's most abundant fisheries region, the North Pacific, for renewed fishing effort where over 90 percent of the fisheries still remained unregulated.[91] Japan was not obliged to adopt restrictive conservation measures and in the two cases where international organizations were established, the INPFC and the Soviet-Japan Fisheries Commission, the quota-making body was liberal in its application and largely served to open up abundant fishery zones for Japanese use.

Consequently, Japan was largely successful in its goal of attaining a measure of food security by securing access to overseas fishing grounds for its distant-water fleets. This, in turn, allowed for a high degree of self-sufficiency in fisheries production that would continue into the 1970s. Increasing environmental concerns and the resource nationalism of coastal states, however, would play an important role in the rise of an enclosure movement during the 1970s that challenged the freedom of the seas regime by calling for greater restrictions on fishing in the high seas and expanded control of fishing grounds in water contiguous to coastal nations. The rise of the coastal state enclosure movement, its subsidiary application of increasing restrictions on high seas fishing and the impact these movements had upon Japan's motives for international fisheries policy are the subjects of the next chapter.

Notes

1 Part of this chapter is based upon a previously published article, Roger D. Smith, 'Food Security and International Fisheries Policy in Japan's Postwar Planning', *Social Science Japan Journal*, Vol. 11, No. 2, 2008, pp 259–76. Permission granted.
2 Bernard Cohen, *The Political Process and Foreign Policy: The Making of the Japanese Peace Settlement*. Princeton, NJ: Princeton University Press, 1957, 153–79.
3 US Department of State, *Foreign Relations of the United States (FRUS)*, 1951, vol. VI, 1184.
4 Ibid., 1184.
5 Kawakami Kenzō, *Sengo no kokusai gyogyō seido* (The Post-war International Fisheries Regime). Tokyo: Dai Nihon Suisankai, 1972, 148–50, emphasis added.
6 Only a few fisheries were actually managed under conservation systems in 1951: halibut, herring and salmon on the north-west coast of North America; seals in the Arctic; most whales worldwide, primarily in the Antarctic. Japan was actively catching both seals and whales in 1940.
7 *Nihon gaikō shuyō bunsho nenpyō 1* (Basic Documents on Japanese Foreign Relations, vol. 1), Kajima Heiwa Kenkyujo, ed. Tokyo: Hara Shobō, 1984, 419–39.
8 US Ambassador Dulles rejected Australian demands for the prohibition of Japanese fishers in Australian waters by explicitly saying that it would be a 'repugnant precedent' that would result in wider calls for Japan's exclusion from high seas fisheries. Found in Harry Scheiber, *Inter-Allied Conflicts and International Law, 1945–53: The Occupation Command's Revival of Japanese Whaling and Marine Fisheries*. Taipei: Academia Sinica, 2001, 192–93.
9 Ministry of Foreign Affairs (MOFA), *Kita taiheiyō no kōkaigyogyō ni kansuru kokusai jōyaku kankei iken* (Opinions Regarding International Agreements on National Fisheries in the North Pacific). Microfilm Series B'0039, MOFA Archives.
10 MOFA, *Gyogyō kyōtei ni kansuru hōshin* (Policy with Respect to Fisheries Treaties). MOFA Archives, Microfilm Series, 0084-99.
11 Ibid.
12 MOFA, *Tripartite Fisheries Conference, Tokyo, Japan, 1951*. Tokyo: Ministry of Foreign Affairs, 1951, 176.
13 MOFA, *Nichibeika sankoku gyogyō kyōtei kōshō ni kansuru hōshin* (Policy with Respect to the Trilateral Fisheries Negotiations among Japan, the United States and Canada). MOFA Archives, Microfilm Series, 0099-0100.
14 *FRUS*, 1951, vol. XIV, 1446–47.
15 *Pacific Fisherman*, January 1951. Seattle: Miller Freeman Publications: 15.
16 Ibid., 15.
17 *FRUS*, vol. I, 1951, 204–5.
18 The State Department resisted attempts to levy tariffs on Japanese tuna imports in 1951–52 even when American tuna boats were dry-docked due to poor market conditions. Their reasoning, as outlined in a report submitted to a Senate Ways and Means Committee in 1952, highlighted two urgent matters: first, Japanese exports were an important source of dollars for post-war reconstruction; and

second, further trade restrictions might have pressured Japan to trade heavily with communist China and driven the Japanese to pursue unfair trade practices in other products in an effort to earn the dollars they needed. Testimony of Harold Cinder, Secretary of State for Economic Affairs before the Senate Finance Committee, 6 February 1952. Found in MOFA, *Kita taiheiyō no kōkai gyogyō ni kansuru kokusai jōyaku kankei iken.* The development of export markets for Japan also concurs with the American objective to bridge the 'dollar gap' that had developed between the United States and its Allies. See William S. Borden, *The Pacific Alliance: United States Foreign Economic Policy and Japanese Trade Recovery, 1947–1955.* Wisconsin: University of Wisconsin Press, 1984, 18–60.

19 MOFA, *Tripartite Fisheries Conference*, 16–17.

20 International North Pacific Fisheries Commission (INPFC), *International North Pacific Fisheries Commission Handbook.* Vancouver: INPFC, 1990.

21 INPFC, *Handbook*, passim.

22 Kaminaga Eisuke, 'Hokuyō to wa nanika, saikō chikusareta gyogyōshi to tairo-kan' (Defining the 'Northern Seas', Opinions on the Reconstruction of Fisheries History with Respect to Russia), in *Roshia no naka no ajia/ajia no naka no roshia I* (Asia in Russia/Russia in Asia 1). Sapporo: Hokkaido University Slavic Research Centre, 2004, 78–79.

23 Kimie Hara, *Japanese-Soviet/Russian Relations since 1945.* London: Routledge Publishers, 1998, 15–20.

24 Savitri Vishwanathan, *Normalization of Japanese-Soviet Relations, 1945–1970.* Tallahassee, FL: The Diplomatic Press, 1973, 113.

25 Ibid., 113.

26 Kawakami, *Sengo no kokusai gyogyō seido*, 412–14.

27 Vishwanathan, *Normalization of Japanese-Soviet Relations*, 115.

28 Ibid., 114.

29 Fisheries Agency, *Suisanchō 50-nenshi* (Fifty-Year History of the Fisheries Agency). Tokyo: Suisanchō 50-nenshi kyōkai, 1998, 101–2.

30 Ibid., 102.

31 Kawakami, *Sengo no kokusai gyogyō seido*, 411–12.

32 Vishwanathan, *Normalization of Japanese-Soviet Relations*, 116.

33 Derek da Cunha, *Soviet Naval Power in the Pacific.* Boulder, CO: L. Rienner Publishers, 1990, 46–47.

34 Ibid., 76.

35 Kawakami, *Sengo no kokusai gyogyō seido*, 415.

36 Hara, *Japanese-Soviet/Russian Relations since 1945*, 63.

37 Ibid., 62.

38 Fisheries Agency, *Suisanchō 50 nenshi*, 103.

39 Donald Hellmann, *Japanese Domestic Politics and Foreign Policy: The Peace Agreement with the Soviet Union.* Berkeley, CA: University of California Press, 1969, 176–79.

40 Fisheries Agency, *Suisan-cho 50-nen shi*, 103.

41 *Nihon gaikō shuyō bunsho nenpyō 1*, 755–56.

42 Ibid., passim.

43 MOFA, 'Hokusei taiheiyō nisso gyogyō iinkai daiikkai kaigi no gijiroku' (Records of the First Meeting of the Committee for Northwest Pacific Soviet-Japanese Fisheries), in *Gaikō Seisho (I)* (Diplomatic Blue Book I). Tokyo: Ministry of Foreign Affairs, 1957, 182–94.

44 Takashi Inoguchi and Nobuharu Miyatake, 'Negotiation as Quasi-Budgeting: The Salmon Catch Negotiations between Two World Fisheries Powers', *International Organization* vol. 33 (Spring 1979): 240–41.

45 INPFC, *Historical Catch Statistics for Salmon of the North Pacific Ocean*, Bulletin 39. Vancouver: INFC, 1979.

46 Yoshihide Soeya, *Japan's Economic Diplomacy with China, 1945–1978*. Oxford: Oxford University Press, 1999, 20–42.
47 Shigeru Oda, 'Nitchū gyogyō kyōtei no seiritsu o megutte' (Some Comments on the Conclusion of the Japan-China Fisheries Agreement), *Juristo* no. 84 (15 June 1955): 32.
48 It should be recalled that the PRC was still engaged in the Korean War against the United States and its allies at this time. Since Japan was providing logistical and material support to American troops in the war effort, it is possible that the PRC's seizure of Japanese vessels and severe application of regulations were designed to undermine this important US ally.
49 Zengo Ohira and Terumichi Kuwahara, 'Fishery Problems between Japan and the People's Republic of China', *The Japanese Annual of International Law* no. 3 (1959): 111.
50 Oda, 'Nitchū gyogyō kyōtei no seiritsu o megutte', 30.
51 Fisheries Agency, *Suisanchō 50 nenshi*, 111.
52 Ibid., 111.
53 Ohira and Kuwahara, 'Fishery Problems between Japan and the People's Republic of China', 114.
54 The Japanese delegation was composed of the Dai Nippon Maritime Industry Association, the Japan Pelagic Trawl Fishery Association, the PRC-Japanese Fishing Enterprisers' Association, the Japan Fishery Combined Cooperative Association, the Japan Seamen's Union, the Japan Marine Products Refrigeration Labour Union Council, and the Marine Products Research Society. Fisheries Agency, *Suisanchō 50-nen shi*, 111–12.
55 *Nitchū kankei kihon shiryōshū 1947–71* (Collection of Basic Documents on Japanese-PRC Relations 1945–71). Tokyo: Nitchū Kokkyō Kaifuku Sokushin Giin Renmei, 1971, 325–34.
56 Ohira and Kuwahara, 'Fishery Problems between Japan and the People's Republic of China', 120–21.
57 Choon-ho Park, 'Fishing under Troubled Waters: The Northeast Asia Fisheries Controversy', *Ocean Development and International Law* vol. 1, no. 2 (1974): 116.
58 Ibid., 114–15. The Rhee Line demarcation will be discussed later in this chapter.
59 *Nitchū kankei kihon shiryōshū 1947–71*, 325–34.
60 Ibid., 325–34.
61 Fisheries Agency, *Suisanchō 50-nenshi*, 115.
62 Masahiro Miyoshi, 'New Japan-China Fisheries Agreement: An Evaluation from the Point of View of Dispute Settlement', *The Japanese Annual of International Law* vol. 41 (1998): 32.
63 Ibid., 32.
64 Ohira and Kuwahara, 'Fishery Problems between Japan and the People's Republic of China', 121–22. It should also be mentioned that many vessels were seized during this time as a result of Chinese frustration with the lack of enforcement of agreed rules by Japanese authorities. For example, the Fisheries Agency found 196 vessels in violation of the agreement between April 1955 and December 1957, but only forty such offenders were punished. See Park, 'Fishing under Troubled Waters', 116.
65 *Seikei bunri* is a term closely associated with Prime Minster Ikeda's foreign relations philosophy and is derived from the Chinese characters meaning 'separation of politics and economics'.
66 Zou Keyuan, 'Sino-Japanese Joint Fishery Management in the East China Sea', *Marine Policy* vol. 27 (March 2003): 126–27.
67 Park, 'Fishing under Troubled Waters', 101–2.
68 Ibid., 102.
69 See, for example, *FRUS*, 1951, vol. VI, 1182–83.

70 Shinji Taguchi, 'Japanese-Korean Fishery Dispute', *Contemporary Japan* vol. 22, no. 7–9 (1953): 397.
71 Kawakami, *Sengo no kokusai gyogyō seido*, 238.
72 Hideo Takabayashi, 'Normalization of Relations between Japan and the Republic of Korea: Agreement on Fisheries', *Japanese Annual of International Law* vol. 10 (1966): 16.
73 Kawakami, *Sengo no kokusai gyogyō seido*, 238.
74 Fisheries Agency, *Suisanchō 50 nenshi*, 104.
75 The full text of the Rhee Line declaration (in Japanese) can be found in *Nihon gaikō shuyō bunsho nenpyō 1*, 471–72.
76 Takabayashi, 'Normalization of Relations between Japan and the Republic of Korea', 82.
77 *FRUS*, 1952–54, vol. XIV, 1486–87.
78 Ibid., 1486–87.
79 *Nihon gaikō shuyō bunsho nenpyō 1*, 419–40.
80 Kawakami, *Sengo no kokusai gyogyō seido*, 239–40.
81 Fisheries Agency, *Suisanchō 50 nenshi*, 105–6.
82 Ibid., 105–6.
83 Kawakami, *Sengo no kokusai gyogyō seido*, 239–50.
84 Park, 'Fishing under Troubled Waters', 103.
85 J.H. Lee, *Korean-Japanese Relations: The Process of Diplomatic Normalization, 1951–1965*. Unpublished thesis, University of Oxford, 183–85.
86 *Nihon gaikō shuyō bunsho nenpyō 2*, 572–75.
87 Ibid., 572–75.
88 Ibid., 572–75.
89 Ibid., 572–75.
90 Fisheries Agency, *Suisanchō 50 nenshi*, 106.
91 Hiroshi Kasahara and William Burke, *North Pacific Fisheries Management*. Washington, DC: Resources for the Future, 1973, 43.

Bibliography

Cohen, Bernard. *The Political Process and Foreign Policy: The Making of the Japanese Peace Settlement*. Princeton, NJ: Princeton University Press, 1957.
da Cunha, Derek. *Soviet Naval Power in the Pacific*. Boulder, CO: L. Rienner Publishers, 1990.
Hara, Kimie. *Japanese-Soviet/Russian Relations since 1945*. London: Routledge Publishers, 1998.
Hellmann, Donald. *Japanese Domestic Politics and Foreign Policy: The Peace Agreement with the Soviet Union*. Berkeley, CA: University of California Press, 1969.
Inoguchi, Takashi and Nobuharu Miyatake. 'Negotiation as Quasi-Budgeting: The Salmon Catch Negotiations between Two World Fisheries Powers', *International Organization* vol. 33 (Spring 1979): 229–56.
INPFC (International North Pacific Fisheries Commission). *International North Pacific Fisheries Commission Handbook*. Vancouver: INPFC, 1990.
Kajima Heiwa Kenkyujo, ed. *Nihon gaikō shuyō bunsho nenpyō 1,2* (Basic Documents on Japanese Foreign Relations, vols 1 and 2). Tokyo: Hara Shobō, 1984.
Kaminaga Eisuke, 'Hokuyō to wa nanika, saikō chikusareta gyogyōshi to tairokan' (Defining the Northern Seas: Opinions on the Reconstruction of Fisheries History with Respect to Russia), *Roshia no naka no ajia/ajia no naka no roshia I* (Asia in

Russia/Russia in Asia 1). Sapporo: Hokkaido University Slavic Research Centre, 2004.

Kasahara, Hiroshi and William Burke. *North Pacific Fisheries Management*. Washington: Resources for the Future, 1973.

Kawakami Kenzō. *Sengo no kokusai gyogyō seido* (The Post-war International Fisheries Regime). Tokyo: Dai Nihon Suisankai, 1972.

Ministry of Foreign Affairs. *Tripartite Fisheries Conference, Tokyo, Japan, 1951*. Tokyo: Ministry of Foreign Affairs, 1951.

Park, Choon-ho. 'Fishing Under Troubled Waters: The Northeast Asia Fisheries Controversy', *Ocean Development and International Law* vol. 1, no. 2 (1974): 93–135.

Scheiber, Harry. *Inter-Allied Conflicts and International Law, 1945–53: The Occupation Command's Revival of Japanese Whaling and Marine Fisheries*. Taipei: Academia Sinica, 2001.

Smith, Roger D. 'Food Security and International Fisheries Policy in Japan's Postwar Planning', *Social Science Japan Journal* vol. 11, no. 2, 2008.

Soeya, Yoshihide. *Japan's Economic Diplomacy with China, 1945–1978*. Oxford: Oxford University Press, 1999.

Vishwanathan, Savitri. *Normalization of Japanese-Soviet Relations, 1945–1970*. Tallahassee, FL: The Diplomatic Press, 1973.

4 The worldwide enclosure movement and restrictive regime claims on fisheries[1]

The purpose of this chapter is to explain how Japanese distant-water fisheries faced an increasingly restrictive international maritime legal regime that placed limits, and in some cases total moratoriums, on Japanese fishing activities in international waters from the 1970s onward. These restrictions developed as a result of changing international practices and rules that gave coastal states a greater say in fisheries resource management decisions internationally.

The post-war bilateral fisheries treaty system that Japan had negotiated in the early post-war years and the internationally accepted ocean regime of freedom of the sea enabled Japanese fisheries to expand their operations far and wide. Japanese fleets had open access to highly valued migratory species on the high seas and to the most productive fishing grounds near the coasts of other states as well. The prevailing legal system of freedom of the high seas facilitated this expansion and enabled Japanese policy makers to achieve their goal of 'from coastal, to offshore, to distant-water fisheries'. Japanese distant-water fisheries were dependent upon the principle of freedom of the seas for secure access to maritime resources.

In the 1960s and 1970s, however, a movement to restrict access to coastal fisheries and place limits on the freedom of the high seas raised new challenges to Japanese distant-water operations. The 'enclosure movement' represented a worldwide trend whereby coastal states claimed ownership of fishing grounds on the continental shelves adjacent to their territories. The enclosure movement sought to change international law through the newly convened United Nations Convention on the Law of the Sea (UNCLOS) to provide coastal states with exclusive management decisions and control of fisheries resources found in waters contiguous to their borders. By the mid-1970s, exclusive economic and fishing zones extending up to 200 nm from the coastline became accepted international practice and as many states, including the United States and the USSR, claimed such exclusive fisheries zones, Japan found itself excluded from many of its most valuable distant-water grounds. The Japanese goal of securing access to a stable fish supply was thus undermined by a new system that restricted freedom of the high seas and extended coastal state control over the world's greatest fishing grounds.

Coastal state claims to these exclusive economic zones (EEZs) were not the only threat posed to Japanese distant-water fishing interests. Several new international movements developed in the 1970–90 period which also placed rather severe restrictions on Japanese fishing activities. The first of these was a concerted effort by conservationist-minded states and environmental groups to put an end to whaling, in the International Whaling Commission (IWC) in the late 1970s and early 1980s. The whaling moratorium that was finally passed in 1982 was the first such measure ever taken to place a zero catch quota on a high seas fishery. Japanese industry and government leaders saw this as a dangerous precedent that could eventually spread to other much more valuable fisheries in the future. Similar controversies arose over the use of driftnets in the North and South Pacific, fishing in the Bering Sea 'Doughnut Hole', as well as salmon fishing in the North Pacific, whereby environmentalists and coastal states' interests successfully lobbied to have these fisheries banned as well. Debates surrounding the scarcity of tuna resources also sparked intense exchanges in CITES (Convention on International Trade in Endangered Species of Wild Fauna and Flora) conferences, where Japan won a narrow victory to continue fishing and trading for Northern bluefin tuna species.

The trend of increasing restrictions applied to Japanese distant-water fishing was a harsh lesson for the Japanese. Given that Japan sees itself as a maritime nation, many leaders in industry and the government felt that their fishing industry was under siege by international interests and their goal of secured access to fishery resources was under threat. Japan's goal of food security has been substantially undermined by the widespread adoption of restrictions and moratoriums worldwide. This chapter will explain why and how the international regime for fisheries shifted its support from the principle of freedom of the seas to coastal state stewardship of marine resources. Japanese officials' responses, in terms of motives and means of foreign policy, to this increasingly restrictive application of international rules will then be investigated in the next chapter.

Normative shifts and international codification: the enclosure movement and UNCLOS

Japan's distant-water fishing industry prospered under the bilateral fisheries regime that Japan negotiated in the North Pacific in the early post-war period. Driftnet fishing helped substantially to increase, albeit indiscriminately, the catch of fish such as squid, salmon and herring. As indicated in Chapter 1, the high seas fishing industry surpassed both coastal and off-shore fisheries production by the end of the 1960s. The total production figure for Japan in 1975, 10.5 million tons of fish, represented a 150 percent increase from 1962 records, and of that total, 83 percent was caught in the North

Pacific including oceans surrounding Japan.[2] Consequently, Japan favoured the continuance of the freedom of the seas status quo as it entered into its third decade of post-war fisheries.

Concurrent developments concerning a new approach to oceans law, however, threatened to undermine Japan's increased access to living ocean resources. In particular, leading fisheries scientists were raising questions with regard to the biological limits to growth – irrespective of the operating needs of industry – which prompted a renewed debate over the sustainability of fisheries and the provision of adequate international regulatory controls to protect fishing grounds.[3] In many cases it was found that international law concerning fisheries was inadequate, ambiguous and lacked enforcement and appeal processes.

After years of speculative debate, the UN General Assembly finally sought in the late 1950s to codify the law of the seas in light of changing practices and policies. The establishment of 200-nm 'conservation zones' off the west coast of South America in a multilateral arrangement by Chile, Ecuador and Peru in 1952 and the unilateral extension of the territorial sea for nine states by 1944 are just two examples of the challenges posed to the old regime in the early post-war period.[4] As a result, the UN elected to host a worldwide multilateral conference to negotiate a new law of the sea to reconcile the conservation concerns of coastal states with the rights and obligations with respect to open access to ocean resources.

The first UNCLOS conference convened in Geneva in 1958. The conference adopted four separate conventions: the Convention on the Territorial Sea and the Contiguous Zone; the Convention on the High Seas; the Convention on Fishing and Conservation of the Living Resources of the High Seas; and the Convention on the Continental Shelf. Although the Convention on the Continental Shelf granted coastal states 'sovereign rights for the purpose of exploring it and exploiting its natural resources', the wording of the articles was otherwise ambiguous and did not adequately define what constituted a continental shelf, nor did they provide an agreed-upon breadth.[5] Moreover, fishing was expressly identified as a protected 'freedom of the sea', and no agreement was achieved with respect to coastal fishing rights for international fishers.[6] Thus, while the 1959 negotiations may have indicated a gradual shift in values with respect to coastal state claims on ocean resources, the conventions were not yet codified as international law due to lack of consensus on crucial issues.

A second conference (UNCLOS II) was held a year later, in 1960, to resolve these jurisdictional conflicts. A North American proposal for a six-mile territorial sea and fishing zone attracted a great deal of support, but was finally defeated by a single vote.[7] The conference adjourned without the resolution of any of the significant issues. Again, no limits for the zones of coastal-state jurisdiction were agreed upon or included in the conventions. Consequently, the old 3-mile territorial sea and the principle of freedom of the sea continued as de facto international custom.

These two UNCLOS sessions reflected a movement to protect coastal fisheries that was gaining momentum around the world. Increased technological capacity and demand intensified the pressures on living maritime resources worldwide. Within the North Pacific, as in other oceans worldwide, coastal states were becoming increasingly alarmed at the mounting evidence of over-fishing and other forms of resource depletion that brought to an end any belief in the age-old Grotian premise of the inexhaustibility of the oceans. Correspondingly, coastal states united in an enclosure movement to bring contiguous fishing areas under national sovereign control in the interest of more effective management and conservation practice. By 1974, thirty-three nations had declared extended fisheries jurisdiction beyond 12 miles.[8]

As UNCLOS launched into its third round of talks in 1973, the Japanese government recognized that it needed to address this increasingly influential movement. It became apparent at the first procedural session held in New York in 1973 that special EEZs concerning coastal control of fisheries and other natural resources would dominate the forthcoming debate. The simple advocacy of general freedom of the high seas would no longer suffice to counter increasingly influential coastal state claims of extended jurisdiction.

The Ministry of Foreign Affairs (MOFA) thought it would be prudent of Japan to declare its own EEZ and then negotiate bilaterally with neighbouring countries regarding access agreements. In a talk to the Japan Press Club in May 1974, Director-General Sugihara Shinichi of MOFA's Office for the Law of the Sea Conference declared that the government should abandon opposition to the 200-nm zones and, instead, discuss the options for Japan to adopt its own economic zone.[9] MOFA was particularly concerned that the adoption of EEZs by such states as the People's Republic of China, South Korea or the Soviet Union might impinge upon Japan's control of its own coastal resources. Moreover, it did not wish to find Japan isolated from an international consensus on oceans law.

The Ministry of Agriculture, Forestry and Fisheries (MAFF) and large distant-water fishing companies, however, did not wish to support a concept that might exclude Japan's international fishing fleets from productive overseas fishing areas. The high seas fishing industry firmly opposed the idea of Japan declaring its own fishing zone for fear it might validate the declarations of other coastal states. The industry, consequently, pressured the Fisheries Agency of MAFF to resist the MOFA proposal. As a result of pressure from such groups as the Japan Fisheries Association – of which the president of Taiyo fisheries was a vice-director – the MAFF minister recommended with the Cabinet's formal support to oppose the concept of the 200-nm economic zone at the second UNCLOS III session in Caracas.[10] As a result of these actions, the Japanese delegation to UNCLOS III, which was composed of both MOFA and MAFF officials, remained split with regard to EEZs as it entered the Caracas negotiations in 1974.

The division within the Japanese group became increasingly apparent as the Caracas meeting progressed. One study of Japan's diplomatic style at the

UNCLOS sessions suggested that the Japanese were in disarray and largely ineffectual at Caracas.[11] Much of the confusion stemmed from the inability of the delegates to establish a clear position before the negotiations. MOFA was shocked when it found out that Japan was the only nation at UNCLOS III actively to oppose the 200-nm proposition, largely due to the delegation's incapacity to forge successful coalitions and formulate effective alternatives. Japan's performance at the conference, according to one participant, was characterized by 'silence, smiling, and sleeping'.[12] Although no convention text was entirely agreed upon at Caracas, the 200-nm EEZ concept was widely supported – most notably by the United States.

Hereafter, the Japanese Fisheries Association and MAFF elected to reassess their policy in light of Japan's isolation regarding the fishery zone issue. The fishing industry abandoned its opposition to coastal states' extended jurisdiction in favour of the advocacy of Japan's traditional fishing rights.[13] In subsequent UNCLOS III negotiations, MAFF and MOFA argued that living resources were the common heritage of mankind, even if found within the 200-nm EEZs. Therefore, longstanding fishing countries such as Japan should enjoy access privileges based on traditional rights notwithstanding the new sovereign rights of coastal states.[14] Moreover, the Japanese delegates argued that coastal states were no more capable of conservation or management of living resources than fishing states. Consequently, international organizations, such as the International North Pacific Fisheries Convention (INPFC), should be strengthened and entrusted with such responsibilities.[15]

Thus, whereas the means used by MAFF and MOFA policy were at odds during early UNCLOS III negotiations regarding whether to support 200-nm EEZs, the same policy goal motivated both agencies. Both MAFF and MOFA still wished to retain the greatest possible allocation of fisheries resources permitted under the prevailing international oceans system. Once it was recognized, however, that the enclosure movement had garnered insurmountable support and that Japan might find itself isolated internationally by resisting this movement, international fisheries policy interests unified in support of traditional fishing rights and calls for the declaration of an exclusive fishing zone for Japan.

Unilateral enclosure in the North Pacific

Two other significant developments prompted the Japanese government to reconsider its ocean policy options in the mid-1970s. The declaration of fisheries zones by the United States and the Soviet Union in 1976 and 1977, respectively, forced Japan to rethink its international fisheries strategy on declaring its own exclusive fishing zone.

As a result of the widespread acceptance of EEZs at the UNCLOS meetings, the United States decided in favour of passing the Magnuson Fishery Conservation and Management Act in 1976 to establish its own 200-nm fishery conservation zone. This represented a reversal of US policy regarding freedom of the seas. Previously, Washington had adhered to the principle of *mare liberum* to enable the free navigation of its vessels, largely for security reasons.[16] This policy was finally relinquished in favour of the protection of the 62 percent of the fish species that were 'fully utilized' or 'overfished' within 200 nm of the American coastline – mostly in the Atlantic Ocean.[17] Washington was no longer willing to compromise domestic fishing interests in favour of Cold War policy, which included bolstering its Asian ally, Japan.[18]

The Act established the US Fishery Conservation Zone (FCZ), allowing the United States exclusive management authority over all fish, all anadromous species spawning in the fresh and estuarine waters of the United States, and all living resources of the US continental shelf (highly migratory species such as tuna were not included under US management). The Act authorized the United States to negotiate treaties with other countries to enable access to fishing grounds within the FCZ.

The Soviet Union soon followed suit with the declaration of its own 200-nm fishing zone in March 1977. Much like Japan, the USSR had opposed the movement to enclose the oceans with fisheries zones at all UNCLOS negotiations, largely in an effort to protect the interests of its own sizeable distant-water fishing fleet.[19] Although the Soviet Union initially resisted the US declaration of its own fishing zone, it finally decided to establish its own

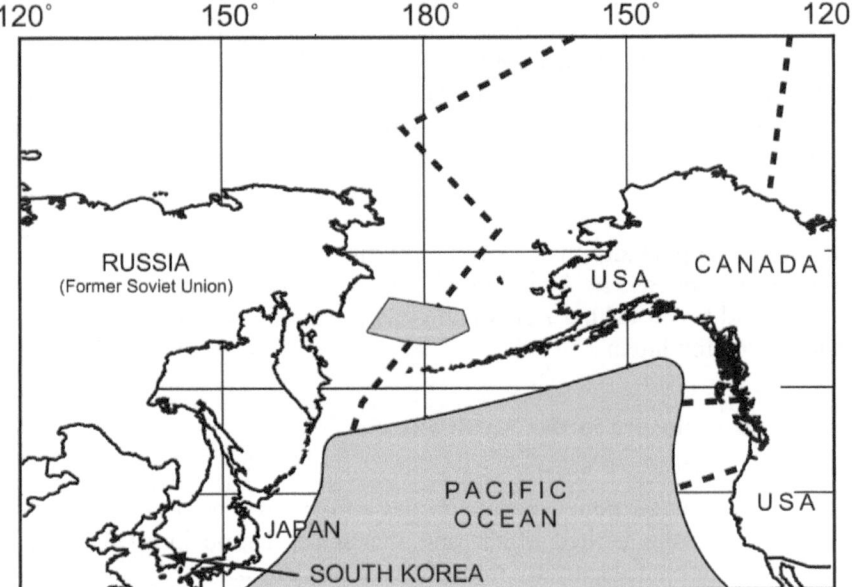

Figure 4.1 EEZs in the North Pacific
Note: Shaded portions represent high seas while non-shaded maritime areas are national EEZs.

fishing zone once the idea of coastal-state enclosure received widespread support at UNCLOS III negotiations.[20] The zone covered waters bordering on all coastal areas, including many traditional Japanese fishing grounds in the North-West Pacific.

As a result of the fisheries claims of the Soviet Union and the United States in the North Pacific, the fishing grounds on which Japan traditionally relied fell within the national jurisdiction of these two coastal states. According to catch statistics for 1975, close to 2.8 million tons of production fell within the fisheries zones of the United States and USSR alone – 27 percent of Japan's total catch.[21]

Japan's fisheries zone

The US and Soviet declarations unilaterally appropriated many of Japan's most valuable fishing areas in the North Pacific under national control by the beginning of 1977. Initially, Japan was reluctant to declare its own fisheries zone for fear of implicitly sanctioning these unilateral declarations of EEZs that deprived Japan of many of its international fishing areas.[22]

The threat of an extended Soviet fishing area, however, prompted coastal fishing interests, particularly in northern Japan, to defend their own domestic fisheries from the encroachment of Soviet fishers with the establishment of a counter-fisheries jurisdiction. Nearly 2,000 claims of illegal fishing were lodged by local fishermen between 1971 and 1977, mostly against Soviet fishing vessels, and nearly half of these incidents occurred within 12 miles of the Japanese coast.[23] Coastal fishermen demanded that the government of Japan extend the territorial sea to 12 miles, as was the custom in many other states around the world, and establish a fishing zone to protect their traditional fishing grounds from foreign encroachment.

Moreover, Japan faced a deadlock in February 1977 in negotiations with the USSR regarding Japanese fishing in newly declared zones near the Soviet coastline. The director of the Fisheries Division within MOFA, Asomura Kuniaki, stated that the Fisheries Agency and MOFA thought it prudent immediately to establish its own 200-nm fishing zone and 12-mile territorial sea as a means to place Japan on more equal footing with the Soviets, who desired reciprocal access to some Japanese fishing grounds.[24]

The Fisheries Agency coordinated a committee with MOFA and the Maritime Safety Agency to draft a new maritime law for Japan in the spring of 1977. By 1 June the Japanese Diet had passed two new laws, to be effective from 1 July 1977, declaring a territorial sea of 12 miles and a provisional fisheries zone.

The territorial sea was defined in accordance with wording adopted by other nations at UNCLOS. It declared sovereign rights to a distance of up to 12 nm with a provision relating to the free passage of innocent ships. The law also established five international straits that allowed for the passage of ships of all nations, military or non-military, whereby Japan declared a 3-mile territorial boundary to either side of the strait.[25]

Unlike Japan's claim to a territorial sea, the fishery zone was designed as a provisional law. The law stated that measures to protect fisheries resources would be made on a temporary basis pending the outcome of the Law of the Sea conference. Moreover, the law made it clear that Japan did not claim sovereign rights over the fishing zone, but only the right to manage fishing within this boundary. Thus, the expression of 'jurisdiction' was used instead of 'sovereign rights' or 'exclusive authority', as commonly claimed by other nations.[26]

The most interesting feature of this new law was that whereas it laid claim to a fishery jurisdiction to the east and north of the Japanese isles, the fishing zone did not extend into waters contiguous to South Korea or the People's Republic of China (PRC). It exempted nationals of these countries from the application of the law and also excluded significant fishing grounds in the Yellow Sea, the western part of the Japan Sea, the East China Sea and waters adjacent to it in the Pacific Ocean.[27] Since neither the PRC nor South Korea made any indication of an intention to declare their own fishing zones, the Japanese government wished to maintain the status quo in their fishery relations.

Conclusion of UNCLOS III in 1982 and the worldwide adoption of EEZs

Coastal state claims to extended fisheries zones achieved widespread acceptance by the end of the 1970s. Whereas only sixteen countries had declared 200-nm fishing or exclusive economic zones in 1975, ninety-one countries had made such claims by 1982.[28] Almost half of Japan's total catch came from fisheries that fell within the 200-nm zones of other states and 90 percent of this amount originated in the North Pacific.[29] Moreover, UNCLOS was concluded on 10 December 1982, ushering in a new global treaty for the management of the oceans. Japan was facing a significantly changed international oceans regime that deprived it of much of the free access to living marine resources once guaranteed under freedom of the seas.

The most significant provision of the new UNCLOS treaty was the one pertaining to EEZs. Article 56 stated that coastal states have:

> sovereign rights for the purpose of exploring and exploiting, conserving and managing the natural resources, whether living or non-living, of the waters superjacent to the seabed and of the seabed and its subsoil, and with regard to other activities for the economic exploitation and exploration of the zone, such as the production of energy from the water, currents and winds.[30]

Article 57 then permitted contracting parties to claim an EEZ extending up to 200 nm from a coastal state's baseline.[31] Thus, UNCLOS gave coastal states final authority to choose how resources found within its zone were utilized and to whom access and use of such resources were granted, despite the obligation of coastal states to consult other countries.

Another provision of particular concern to Japanese distant-water fisheries relates to the coastal state's authority to determine allowable fish catches within its EEZ. Each coastal state was charged with the obligation to protect and conserve fish stocks found within its EEZ and to determine allowable catch sizes based upon the best scientific evidence possible.[32] The coastal state had the right not only to harvest that catch, but also to determine whether other states were to be given access to any surpluses above and beyond what domestic fisheries were able to harvest. With regard to access to such surpluses, Article 62(3) declared that 'the coastal State shall take into account all relevant factors, including, *inter alia*, the significance of the living resources of the area to the economy of the coastal State concerned and its other national interests ... and the need to minimize economic dislocation in States whose nationals have habitually fished in the zone or which have made substantial efforts in research and identification of stocks'.[33] The requirement to 'minimize economic dislocation' to habitual fishers, such as the Japanese distant-water fleets in many coastal waters around the world, is an acknowledgement of Japan's concerns to protect its fisheries interests through claims of preferential and traditional rights. In terms of levels of priority access to surplus stocks, however, coastal state interests were paramount and habitual fishing rights were the lowest.[34]

Japan was thus obliged under international law to negotiate with coastal states for access to fishing grounds found within other countries' EEZs. Even though its rights were acknowledged and given some protection under the UNCLOS treaty, they were deemed significantly less important than the rights of coastal states to protect fish stocks and determine sustainable levels of harvest for coastal fishermen.[35]

As was mentioned previously, the declaration of 200-nm fishing zones by the United States and the USSR placed many Japanese distant-water fishing grounds under the national control of these governments. As a country with habitual fishing rights in these areas, Japan was entitled to a portion of the surplus allotment of the fishery harvest. Tokyo's negotiating position, however, was substantially weaker than it had been during previous negotiations held at the peak of the Cold War, now that coastal states were given priority access and preferential management rights over coastal resources under international law.

The United States permitted Japanese vessels access to its coastal waters in the years following the declaration of its fishing zone, but the share of harvest was substantially reduced from year to year. In December 1980, the US Senate passed a US fisheries promotion law that aimed to 'Americanize' fishing within the US 200-nm fishing zone by phasing out quota allocations to foreign fishermen and setting up new American fishing enterprises.[36] Table 4.1 outlines the plan adopted by the US government for the Americanization of the trawling fishery in Alaska starting in 1980, which serves as a typical example of how the phase-out was accomplished. The plan called for the complete elimination of foreign trawling by 1990 along with their replacement by Alaskan fishing vessels that were permitted to establish some joint ventures with foreign, mostly Japanese, companies to facilitate market access.

As a result of the Americanization of fishing within the US fishery zone, the United States overtook Japan as the world's leading exporter of fishery products and Japan became increasingly dependent upon American seafood imports. Washington's declaration of a 200-nm zone helped transform the United States from a minor seafood exporter, principally involved in fish oil, to the world's leader specializing in high-valued fresh-frozen fishery products.[37] Meanwhile, the strong demand for high-quality seafood in conjunction with the increased buying power of an appreciated Yen in the mid-1980s in Japan created complementary economic conditions supporting the import of American fresh-frozen seafood. The result of these factors meant that whereas seafood only comprised 8 percent of the value of all US exports to Japan in 1977, this figure jumped to 20 percent by 1990.[38]

Similar to the case of Japanese fishing within the US EEZ, the Soviet Union significantly cut harvesting levels and raised fishing access fees for Japanese fishermen for two decades following the declaration of their own fishing zone. Within the first year that the Soviet fishing zone and provisional Japan-Soviet fishing treaty came into effect in 1977, Japan's fisheries quota in Soviet waters was reduced by half: Alaskan pollack and flatfish were reduced by half to 500,000 tonnes, sardine catches were reduced to zero, salmon

Table 4.1 Fisheries promotion plan for Alaskan trawl fisheries

	Foreign fishing vessels	American fishing vessels	Joint ventures (American trawlers and foreign processing vessels)
1979	97.5	1	1.5
1980	94	3	3
1981	89	5	6
1982	80	10	10
1983	70	20	10
1984	60	30	10
1985	55	35	10
1986	50	40	10
1987	40	50	10
1988	30	60	10
1989	15	75	10
1990	0	90	10

Source: Morikawa Akira, 'Wagakuni enyō-gyogyō no shukushō to sono taiō' (The Contraction of Japanese Distant Water Fishing and Japan's Reaction), in Taki Yasuhiko (ed.) *Sekai no naka no nihon gyogyō* (Japanese Fisheries in the World). Tokyo: Seisando Publishing, 1993, 35

fishing within Soviet waters was banned, while high seas salmon catches (for those salmon originating from rivers within the USSR) was limited to 62,000 tonnes.[39] This trend of reduced fish quotas for Japan continued for the following two decades, as can be seen in Table 4.2, featuring statistics for salmon. Japan's share of the salmon catch fell from the originally halved figure of 62,000 in 1977 to 42,500 tonnes in 1978, whereupon it was further halved again within ten years to 17,668 tonnes in 1988. Meanwhile, the fees for salmon catches were steadily rising despite the reduction in quotas. Salmon fees more than doubled between 1978 and 1983 despite no change in the fisheries quota. Even in 1991, a sizeable 28,400 million yen was paid for a reduced quota of 9,000 tonnes of salmon.

Moreover, Soviet fishermen started to develop more intensely the fishing grounds that were previously exploited by Japanese fishing vessels. Much like the case of the US fishing zone, Japanese fishing efforts were replaced by their Soviet counterparts in the Soviet EEZ.

The declaration of national fishing zones by the United States and the USSR was met with considerable apprehension among Japanese policy makers. As was morosely noted in the Fisheries Agency's White Paper for 1976, 'As a fishing nation with a high level of dependence upon the fisheries of the North Pacific, Japan and its fishermen are entering an era of great difficulty'.[40] The White Paper went on, however, to reaffirm Japan's commitment 'actively' to negotiate new treaties under the new oceans framework.[41] The prospects of retaining valued overseas fishing grounds must have looked grim indeed, in spite of Japan's adherence to the new oceans law. Subsequent

Table 4.2 Japanese quotas and fishing fees for salmon originating from rivers of the USSR

	Fisheries quota (tonnes)	*Fees (hundred million yen)*
1978	42,500	17.6
1979	42,500	32.5
1980	42,500	37.5
1981	42,500	40.0
1982	42,500	40.0
1983	42,500	42.5
1984	40,000	42.5
1985	37,600	42.5
1986	24,500	35.0
1987	24,500	37.0
1988	17,668	33.5
1989	15,000	33.5
1990	11,000	31.5
1991	9,000	28.4

Source: Morikawa Akira, 'Wagakuni enyō-gyogyō no shukushō to sono taiō' (The Contraction of Japanese Distant Water Fishing and Japan's Reaction), in Taki Yasuhiko (ed.) *Sekai no naka no nihon gyogyō* (Japanese Fisheries in the World). Tokyo: Seisando Publishing, 1993, 54

negotiations on oceans law, however, only served to place even greater restraints upon Japanese distant-water fishing.

Conclusion

Over the course of the post-war era, Japan continuously fought for the preservation of the principle of freedom of the seas, in resistance to littoral state appropriation of coastal fishery resources. For over two decades, MOFA and MAFF successfully managed to secure rights of access to important Pacific fishing grounds and elsewhere through various bilateral fisheries agreements negotiated with the United States, Canada, South Korea, the Soviet Union and the PRC, as seen in the last chapter. This was a remarkable achievement for Japan given the context of a generally subordinate role in its political relationship with the United States during the Cold War. Washington, too, played a substantial role in the development of Japan's high seas fisheries by fostering their growth in the service of economic renewal and the maintenance of the mutual security arrangement in Asia. In this role, the United States assisted the food security needs of an important Pacific ally.

As the 1970s neared, however, coastal states recognized the failure of such agreements to mitigate fish-stock depletion. Japan could no longer rely upon American Cold War imperatives to support the increasingly untenable principle of freedom of the seas. The original *mare liberum* regime was now being widely questioned and alternative regimes were being considered. As both the number of fishers and the efficiency of fishing technology increased, the ecological pressures on stocks of living resources magnified and sustainable management of common resources became a widespread priority. Consequently, a movement to introduce national control through expanded state jurisdictions established the current system of EEZs.

Enclosure served to appropriate international resources for domestic exploitation and benefit. The enclosure movement was predicated upon the belief that national control of common resources could be managed to apply more effectively disincentives to overfishing and incentives for conservation. It was hoped that national property rights and rules could circumvent the tragedy of the commons through the creation of domestic vested interests and licensing systems. In consideration, however, of the mounting difficulties encountered by contemporary national fishery departments around the world, the feasibility of the national-control alternative to solve the commons resource problem remains in doubt.

It should be noted that the enclosure movement was initiated as a legalistic and scientific apportionment of fisheries effort. It was not until the 1972 Conference on the Human Environment that ocean issues such as whaling and overfishing became widely known and popularized in Western countries. Hereafter, calls for protection of life in the oceans became a formidable force in ocean politics and international conventions.

Japan lost many of its traditional distant-water fishing grounds due to the enclosure movement. The North Pacific Ocean and its regional seas provided a very important source of fish supply for domestic Japanese consumption and, with the onset of unilateral declarations of exclusive fishing zones – most importantly those of the United States and USSR – Japan was required to reassess its fisheries strategy and declare its own fisheries zone in 1977 in order to protect the interests of domestic fishermen. The fact that the Japanese government repeatedly voiced its opposition to extended fishing zones at UNCLOS negotiations and held out for so long against declaring its own fishing zone is an indication of the influence that the distant-water fishing industry had over governmental agencies.

Japan was also forced to reconcile new international pressures to place restrictions on other distant-water fishing activities. Perhaps of greatest concern is a growing trend to apply moratoriums on fisheries when there are fears of irreparable damage to maritime resources, as was seen in the case of both whaling and driftnets. These moratoriums have the distinctive characteristic of being virtually irreversible once applied, in spite of Japan's best efforts to reverse them. If moratoriums were to be applied more generally to other stocks when endangered, the very basis of Japan's fishery strategy might be under threat.

Another trend worrying to Japanese officials has been the extension of coastal state control into the high seas beyond the allowable 200-nm limits. The Bering Sea 'Doughnut Hole' case, the ban on high seas salmon fishing as well as the Fish Stocks Agreement illustrated how coastal states can claim special interest in management decisions over fish stocks that straddle or migrate into the high seas from their coastal waters. Although Japan has negotiated a place for itself at the table for regional fisheries organizations concerned with the management of such stocks, its claim to habitual fishing rights over these resources is of less influence than the right of coastal states for protection and management. Japan has had to give up, in effect, its right to fish for key stocks such as pollack and salmon on the high seas.

A final trend that is particularly worrying to Japanese officials is the rise of environmental movements and their use of the recently codified and ratified 'precautionary approach'. Pressure from environmental groups and many Western states that have adopted their cause has led to moratoriums on whaling and driftnets, as well as an attempted ban on the trade of bluefin tuna. There are fears that new rallying calls may be made for restrictions on other fisheries, such as deep sea trawling or high seas tuna fishing, and many industry and governmental officials feel that the environmental movement is the single biggest threat to Japan's fishing interests. Environmental groups are also lobbying for the greater application of the precautionary approach to fishing worldwide, which can be used to support bans on fisheries deemed threatened, even in the absence of scientific proof. This is a trend that Japan has tried to counter with the promotion of the scientific-based management of fisheries.

Notes

1 Part of this chapter was previously published in Roger D. Smith, 'Food Security and International Fisheries Policy in Japan's Postwar Planning', *Social Science Japan Journal* vol. 11, no. 2 (2008): 259–76.
2 Food and Agriculture Organization (FAO), *Yearbook of Fisheries Statistics.* Rome: FAO, 1977, c-61, c-67.
3 More can be found on the conservationist/preservationist epistemic debate in Chapter 9.
4 See Douglas Johnson, *The International Law of Fisheries.* New Haven, CT: Yale University Press, 1965, 333–41; and Barry Buzan, *Seabed Politics.* New York: Praeger, 1976.
5 'Convention on the Continental Shelf', *United Nations Treaty Series* vol. 499, no. 7302 (1958): 312.
6 Ibid., 312.
7 James Morell, *The Law of the Sea: An Historical Analysis of the 1982 Treaty and its Rejection by the United States.* Jefferson: McFarland and Co., 1992, 8.
8 Ross Eckert, *The Enclosure of Ocean Resources: Economics and the Law of the Sea.* Stanford: Hoover Institution Press, 1979, 129.
9 *Nihon Keizai Shinbun*, 'Nihyaku kairi hantai sezu' (Japan Will Not Oppose 200nm EEZs), 29 May 1974.
10 *Nihon Keizai Shinbun*, 'Kaiyōhō kaigi: kakushōchō no arasoi hyōmenka-e' (UN Conference on the Law of the Sea: interministerial conflicts rise to the surface), (June 19, 1974).
11 Michael Blaker, 'Evaluating Japanese Diplomatic Performance', *Japan's Foreign Policy After the Cold War: Coping with Change*, Gerald Curtis, ed. New York: M.E. Sharpe, 1993, 13. It should be noted that this author has developed a reputation as a critic of Japan's 'ineffectiveness' at international negotiations and thus his analysis should be considered with due caution.
12 Ibid., 16.
13 'Chōki shokuryō seisaku no ikkan de kakuritsu o: kaiyōhō e gyōkai hōshin, jisseki kakuho no sesshō nado sanken' (Let's Establish Policy as Part of a Long-term Food Policy: The Industries Position on the Law of the Sea, Three Points Including Negotiations for Ensuring Past Fish Catch Levels), *Suisankai* no. 1084 (April 1975): 52–54.
14 Kazuomi Ouchi, 'A Perspective on Japan's Struggle for its Traditional Rights of the Ocean', *Ocean Development and International Law* vol. 5, no. 1 (1978): 121.
15 Ibid., 118.
16 Morell, *The Law of the Sea*, 56.
17 Ross Eckert, *The Enclosure of Ocean Resources: Economics and the Law of the Sea.* Stanford: Hoover Institution Press, 1979, 134. The law was amended in 1980 by the American Fisheries Promotion Act to include a mandatory reduction of foreign within the FCZ that could abolish all foreign fishing by 1990. See Cameron Crone Bilger, 'US-Soviet Fishing Agreement: Treaty Authorizing Soviet Fishing in US Waters', *Marine Policy* vol. 10 (January 1986): 53.
18 It is likely that the US-Soviet détente that emerged in the early to mid-1970s helped the United States in its reassessment of ocean policy by taking off some of the earlier pressure to pursue Cold War strategies.
19 Erik Franck, 'New Soviet Fishery Regulations Concerning the EEZ: An Appraisal', *Marine Policy* vol. 11 (April 1987): 125–26.
20 Ibid., 125–26.
21 Fisheries Agency, *Zusetsu gyogyō hakusho, shōwa 52 nendo* (Illustrated Fisheries White Paper, 1977). Tokyo: Nōrin tōkei kyōkai, 1977, 5.

22 Distant water fishing interests were opposed to the growing trend of unilateral extension of fishing zones by foreign states since such claims appropriated fishing grounds used by their fleets.
23 Shunji Yanai and Kuniaki Asomura, 'Japan and the Emerging Order of the Sea: Two Maritime Laws of Japan', *Japanese Annual of International Law* vol. 21 (1977): 49.
24 Ibid., 66.
25 Ibid., 66.
26 Ibid., 77.
27 Ibid., 74.
28 Morikawa Akira, 'Wagakuni enyō-gyogyō no shukushō to sono taiō' (The Contraction of Japanese Distant Water Fishing and Japan's Reaction), in *Sekai no naka no nihon gyogyō* (Japanese Fisheries in the World), Taki Yasuhiko, ed. Tokyo: Seisando Publishing, 1993, 32.
29 Delegation of Japan to the Third United Nations Conference on the Law of the Sea, 'Statement by H.E. Ambassador M. Ogiso, Head of the Japanese Delegation before the Plenary of the Third United Nations Conference on the Law of the Sea', Caracas, 15 July 1974. Cited in Barbara Johnson and Frank Langdon, 'Two Hundred Mile Zones: The Politics of North Pacific Fisheries', *Pacific Affairs* vol. 49 (Spring 1976): 15.
30 United Nations Convention on the Law of the Sea, Agreement Relating to the Implementation of Part XI of the Convention, Article 56, www.un.org/Depts/los/convention_agreements/texts/unclos/closindx.htm.
31 Ibid., Article 57.
32 Ibid., Article 61(1,2)
33 Ibid., Article 62(3).
34 William Burke, *The New International Law of Fisheries: UNCLOS 1982 and Beyond*. Oxford: Clarendon Press, 1994, 78.
35 Although UNCLOS was not put into effect until 14 November 1994, the Convention had widespread acceptance as customary international law during the period between first signatures and the date of effectuation, and its provisions were tried and tested in international courts in the meantime.
36 Morikawa, 'Wagakuni enyō-gyogyō no shukushō to sono taiō', 34.
37 John T. Sproul, 'Effects of North Pacific 200-Mile Exclusive Economic Zone Management Policy on Japanese Seafood Production, Trade and Food Security', *Bulletin of the Faculty of Fisheries, Hokkaido University* vol. 43, no. 3 (1992): 142.
38 Ibid., 149.
39 Morikawa, 'Wagakuni enyō-gyogyō no shukushō to sono taiō', 54.
40 Fisheries Agency, *Gyogyō hakusho, shōwa 51 nendo* (Fisheries White Paper, 1976). Tokyo: Nōrintōkeikyōkai, 1977, 2.
41 Ibid., 2.

Bibliography

Blaker, Michael. 'Evaluating Japanese Diplomatic Performance', in *Japan's Foreign Policy After the Cold War: Coping with Change*, Gerald Curtis, ed. New York: M.E. Sharpe, 1993.
Burke, William. *The New International Law of Fisheries: UNCLOS 1982 and Beyond*. Oxford: Clarendon Press, 1994.
Buzan, Barry. *Seabed Politics*. New York: Praeger, 1976.
Eckert, Ross. *The Enclosure of Ocean Resources: Economics and the Law of the Sea*. Stanford: Hoover Institution Press, 1979.

FAO (Food and Agricultural Organization). *Yearbook of Fisheries Statistics*. Rome: FAO, Series, n.d.

Government of Japan Fisheries Agency. *Zusetsu gyogyō hakusho, shōwa 52 nendo* (Illustrated Fisheries White Paper). Tokyo: Nōrin tōkei kyōkai, 1977.

Johnson, Douglas. *The International Law of Fisheries*. New Haven, CT: Yale University Press, 1965.

Morell, James. *The Law of the Sea: An Historical Analysis of the 1982 Treaty and Its Rejection by the United States*. Jefferson: McFarland and Co., 1992.

Morikawa Akira. 'Wagakuni enyō-gyogyō no shukushō to sono taiō' (The Contraction of Japanese Distant Water Fishing and Japan's Reaction), in *Sekai no naka no nihon gyogyō* (Japanese Fisheries in the World), Taki Yasuhiko, ed. Tokyo: Seisando Publishing, 1993.

Ouchi, Kazuomi. 'A Perspective on Japan's Struggle for its Traditional Rights of the Ocean', *Ocean Development and International Law* vol. 5, no. 1 (1978): 107–34.

Sproul, John T. 'Effects of North Pacific 200-Mile Exclusive Economic Zone Management Policy on Japanese Seafood Production, Trade and Food Security', *Bulletin of the Faculty of Fisheries, Hokkaido University* vol. 43, no. 3 (1992): 124–51.

Yanai, Shunji and Kuniaki Asomura. 'Japan and the Emerging Order of the Sea: Two Maritime Laws of Japan', *Japanese Annual of International Law* no. 21 (1977): 48–114.

5 The precautionary principle, EEZs and fisheries enforcement in the Pacific

The enclosure movement was not the only international trend that arose to challenge the notion of freedom of the seas. The 'precautionary principle' – a legalistic notion that suggested that environmental risk should be minimized, akin to the expression 'better safe than sorry' – began to gain currency in the 1960s and impel diplomatic negotiations in favour of resource protection and fisheries restrictions. Concomitantly, several developments occurred that would have the net effect of restricting Japanese fishing activities in international waters.

The most important of these movements included efforts to ban whaling and driftnets, attempts to have bluefin tuna stocks protected as an endangered species, adoption of a moratorium on salmon catches in the Pacific high seas, pressure to stop fishing in the Bering Sea 'Doughnut Hole', and extending protection of straddling stocks and highly migratory fish in international jurisdictions. These moratoriums, precautionary approaches and extensions of coastal state controls will be investigated in turn to see how they further undermined international support for freedom of the sea and Japan's free access to distant-water fisheries. These will be treated subsequently in chronological order.

Precautionary principle

The precautionary principle is one of the most significant developments in normative and codified law to occur in international environmental law in the past century. As a principle that is still not subject to conclusive consensus nor agreed definition, it nonetheless made a fundamental contribution in terms of reversing onus in environmental degradation cases both domestically and internationally.

After considerable debate among scientists and policy makers in the early post-war 1950s and 1960s, the precautionary principle provided scientists working with natural resources with an important conceptual and analytical premise to urge protectionary and conservationist measures, especially when scientific communities were unable to provide conclusive proof (i.e. a degree of uncertainty) to industry to take such action. In other words, the onus for proof to show that long-term effects were in the greatest interest of all was placed upon the industries exploiting a resource, requiring them to prove that such economic extraction did not lead to long-term or irreversible environmental harm.

First used in legal contexts in the United States prior to the Second World War with respect to drug manufacturing and research, the concept was later applied to environmental case law starting in the 1970s. In more common terms, 'precaution' may be understood with the adage 'better safe than sorry'; its legalistic equivalent suggests that uncertainty does not justify inaction on behalf of those who might profit from a resource. In more recent applications of the principle, 'precaution' may be understood as a strategy for managing risk: it requires policy makers to take anticipatory action to avoid uncertain future risks, given that future harm is always uncertain.[1]

The first legalistic application was used with respect to both food and drug production in the United States in the pre-war era. It was generally understood that companies could not market any food or drugs that might potentially bring harm to consumers, and that companies were required to prove that this was the case before they could market their products. This was also enforceable in the courts and, as such, became a widely understood concept at the time largely because it touched the lives of all citizens. The precautionary principle was later applied in environmental pollution cases in Germany and eventually featured at international conferences on the environment in the 1970s in order to address the environmental harm being done by human activity worldwide. Thus, it has been adopted and codified in both domestic and international legal contexts.

There has been no consensus, however, on a precise definition to be applied in all codification situations. It is widely applied with respect to 'risk management' and is generally understood to be composed of three components: 1 uncertainty does not justify inaction; 2 uncertainty justifies action; and 3 the burden of proof is shifted upon those who might take the environmental risks.[2] In short, shifting the onus onto industry reduces the risk of false negatives (inattention to significant risks), but increases the risk of false positives (attention to insignificant risks), but the overall cost of such a risk transfer is not material, but rather an 'opportunity cost' for the economy, whereby potential wealth generation is foregone. Thus, extra preventative measures ensure that irreparable environmental risk is not transferred to future generations for the gain of the present generation.

The concept of precaution may also be understood as distinct from prevention. Prevention is avoidance of a course of action when the risks involved are certain and, therefore, the task of decision makers somewhat easier to assess. Precaution, on the other hand, involves avoidance when risks are 'uncertain', which is almost always the case when dealing with fisheries science and management.[3]

How precaution was applied with respect to various distant-water fisheries and the role of prohibition in scientific assessments will be considered in several cases in this chapter. As an issue that enjoyed popular sympathy and support in Western countries, anti-whaling measures will be the first topic of consideration.

Pressure to ban whaling

Rising environmentalist pressure in the 1970s led to the adoption of a moratorium on all forms of whaling in 1982. Japan vehemently opposed the ban but was bound by the rules of the International Whaling Commission (IWC) to comply with the measure. The moratorium has been in place since and has largely put an end to Japanese commercial whaling, with a limited scientific whaling catch that offers a small supply of whale meat to the domestic market. A more in-depth treatment of this issue will be offered in Chapter 8 in order to investigate how the eventual resolution of this contentious issue has implications for Japan's future international fisheries policies.

Japanese large-scale industrial whaling as we know it today has its roots in the immediate post-war period when Japan expanded its whaling catch to stave off a shortage of food production at home.[4] By 1958, Japan became the world's largest whaling nation in the Antarctic and, in 1962, Japan reached its all-time production peak with a catch of over 300,000 tons of oil and meat.[5]

Whaling around the world, including Japanese production, was monitored by the IWC, which was established under the International Convention for the Regulation of Whaling (ICRW), signed in Cambridge, UK, in 1946. The Commission's principal function is to oversee the 'orderly development of the whaling industry' and to ensure the 'proper conservation of whale stocks' based on scientific findings.[6]

In its early days, the IWC's main responsibility was the so-called orderly development of the industry and it did not function in the preservationist capacity it does today. This was accomplished through the use of the blue whale unit (BWU) as a way to determine an overall catch quota.[7] This ecologically unsound management technique, coupled with a poor state of whaling science, led to catch levels that were too high and placed many populations of whales at risk of overfishing. Whaling managers failed to provide the management guidelines that might have reduced quotas in a timely manner.[8]

In response to the alarming decline of many great whale populations, most famously the blue whale, the worldwide environmentalist movement adopted the whale as its symbol for environmental preservation. 'Save the whale' became a crucial test for environmentalists' ability to put an end to environmental destruction.[9] These environmentalists pushed for a moratorium on all commercial whaling and lobbied their home governments through rallies and media appeals.

They found a powerful ally in the United States. Washington took the lead in several environmental issues, including whaling, in the early 1970s and was successful in passing a declaration in 1972 at the first international conference on environmental issues, the United Nations Conference on the Human Environment, which called for a total ban on whaling for ten years until the IWC got its house in order.[10] A similar proposal, however, subsequently failed in the IWC. Nevertheless, Japan had to defend its whaling industry in the 1970s by winning the support of like-minded nations. Meanwhile, the IWC became a politically charged institution deeply divided between pro-whaling and anti-whaling nations.

By 1982, non-whaling nations outnumbered whaling ones largely because the United States and environmental groups such as Greenpeace recruited new members into the IWC, thus changing the voting patterns within the Commission.[11] The IWC General Council passed a measure in 1982 calling for a moratorium on the catch of all species of whales irrespective of their stock status. The moratorium took effect in the 1986/87 whaling season and would be reviewed under a new management scheme by 1990.[12]

Japan strongly opposed the whaling moratorium. When Japan, Norway and Iceland threatened to exempt themselves from the provision, a right guaranteed under Article V, paragraph 3 of the ICRW, the United States warned that it might apply the Pelly and Packwood-Magnuson Amendments against them.[13] Japan decided to comply with the moratorium due to fears they might lose access to fishing grounds near the West Coast of Alaska.[14]

Although Japan has complied with the whaling moratorium, it made every effort possible to reverse the ban during subsequent IWC meetings. Japanese officials have argued for a resumed hunt of minke whales, the one whale species that Japanese scientists say has returned to healthy numbers in recent years. Japan has not been successful as yet, largely because anti-whaling countries still have a controlling majority in the IWC. Japan also runs a controversial scientific whaling programme that catches close to 400 minke whales a year for scientific study and the whale meat is sold on the domestic market.

The whaling moratorium is the first moratorium on a fishing industry encountered by Japan and it has raised alarm bells for industry and government officials regarding the possibility that other fisheries could go the same way. These officials see the whaling case as an example of the dangers inherent in unbridled environmentalism. They are trying to renew the IWC's faith in science so that a managed fisheries regime, a 'sustainable-use' regime, can be created so as to reverse the moratorium.

The background and issues surrounding the whaling controversy will be examined in greater depth in Chapter 8. The whaling case provides the first challenge raised by the scientific community with regard to the preponderant role of industrial influence on fisheries management and calls for greater scientific decision making in international bodies, such as the IWC. As a result, the response that this provoked from various governments involved in the Commission is very instructive indeed.

Driftnet moratorium, 1991

Driftnets are large, monofilament gillnets that typically have mesh sizes of 160–170 mm, and are 10 metres deep and up to 50 km long. Japanese vessels had used driftnets since the mid-1960s in their operations in the North Pacific high seas in the search for highly valued salmon, tuna and squid stocks, and for tuna in the South Pacific. Driftnets are indiscriminate, yet are an effective way of catching vast quantities of targeted fish at very little cost and effort. In the 1980s driftnetting became a popular method of fishing on the high seas since it cut down on expensive labour costs for fishing vessels and served as a way to ensure high catch rates after Japanese distant-water fishing declined with the imposition of fishing zones in the USSR and the United States.[15]

By the late 1980s, however, these fisheries too came under international scrutiny and pressure, most vocally from an increasingly influential environmental movement based in the United States. Environmental non-government organizations (NGOs), such as Earthwatch and Greenpeace, pointed to the high incidence of by-catch as well as the dangers of 'ghost nets' – driftnets that are lost but still continue to catch marine life – as evidence of the wanton environmental destruction. In the media mêlée that ensued, driftnets were effectively demonized by dubbing them 'walls of death'.[16]

As a result of the lobbying efforts of these environmental groups, as well as Alaskan fishing interests, the US government passed its first piece of legislation concerning driftnet fishing in 1987, entitled the Driftnet Impact Monitoring, Assessment and Control Act. Lobbyists not only focused on the extensive environmental impact on mammals and seabirds, but also claimed that alarming amounts of salmon were being captured in squid driftnets.[17] The act was designed to allow the Secretary of State and the Department of Commerce to negotiate observer and enforcement programmes for Bering Sea fisheries.[18]

The international community also stepped up pressure by supporting a ban on driftnet fisheries in the United Nations (UN). On 22 December 1989, the General Assembly passed UN resolution 44/225, later reaffirmed by UN resolution 46/215, which stated:

> Moratoria should be imposed on all large-scale pelagic driftnet fishing by 30 June 1992, with the understanding that such a measure will not be imposed in a region or, if implemented, can be lifted, should effective conservation and management measures be taken based upon statistically sound analysis to be jointly made by concerned parties of the international community with an interest in the fishery resources of the region, to prevent unacceptable impact of such fishing practices on that region

and to ensure the conservation of the living marine resources of that region.[19]

The resolution called for a 50 percent reduction within a year of all driftnet fleets, with an eventual move toward a total ban, as well as penalties for those countries that did not comply. It should also be noted that the resolution called for 'statistically sound analysis' in order to prevent the politicization of quota allocations. This would turn out to be the most controversial point of the resolution.

Japan argued that criticisms levelled at driftnet fishing were not based on science or fact, and thus not based on 'statistically sound analysis'. As stated by Hayashi Moritaka, the UN Officer for Ocean Affairs and the Law of the Sea, Japan's position was to call upon states to cooperate in regulating drift-net fishing until it could be scientifically proven that such fishing techniques were detrimental to the environment.[20] The US proposal, however, reversed the onus and said that a ban should take effect immediately unless it could be scientifically proven that driftnetting was safe. Their draft resolution stated it thus: 'unless or until it is agreed that the unacceptable impact of such practice can be prevented and that the conservation of the world's living marine resources can be ensured'.[21] This is commonly seen as the first large-scale application of the now well-known concept of the 'precautionary principle', which later gained greater currency at the Earth Summit in 1992 (UN Conference on the Environment and Development, Rio de Janeiro), and was finally codified in the Law of the Seas with the Straddling Stocks agreement, discussed below. Although a compromise was achieved that sent international observers on driftnetting vessels to accumulate scientific data, Japan was reluctant to abide by the proposed moratorium.

In order to coax a reluctant Japan to accept the UN resolution, the US Department of Commerce stepped up pressure to end the use of driftnets by issuing a regulation on 18 September 1991, which stated that fish taken with driftnets on the high seas would be banned from sale in the United States after 1 July 1992.[22] This was largely done in response to the popular Dolphin Protection Consumer Information Act of 1990 with respect to tuna fisheries in the South Pacific, but was also intended to protect valuable Bering Sea salmon stocks that were being captured by Japanese driftnet fisheries. It provided a strong economic tool to enforce compliance with US wishes.

Japan responded to these foreign pressures by adopting a voluntary ban on driftnet use. As of 31 December 1992, the Fisheries Agency stopped issuing fishing licences to driftnet vessels. Although the government offered various compensation packages to those fishers adversely affected, they were not given licences in other fisheries, such as trawling or purse-seine, since these fisheries were already operating at full capacity.[23]

The UN resolution served as another blow to an already dwindling high seas fishing industry. It is widely believed among Japanese policy makers and fisheries managers that Japan was unfairly attacked by environmental organizations and unduly pressured by the United States to accept a moratorium on driftnet fishing.[24] They argue that politics, and not science, was the motivating force that impelled Japan to abandon a valuable fishery.[25]

In the case of both driftnets and whaling, Japanese officials tend to see themselves as the victims of North American environmentalism. In neither case has there been the kind of popular support or sentiment for environmentalism in Japan that was present in North America and Europe, and which led to these bans. Much like the imposition of 200-nm fishery zones, it was external rather than internal pressure that brought about readjustments that dislocated fishing communities and tended to hurt Japan's international reputation. As a result, Japanese bureaucrats and industry officials tend to view foreign conservationists as a new enemy and a threat to their traditional way of life.[26]

Attempts to list tuna for protection with CITES

Another environmental issue that has raised an alarm within the Japanese fishing industry has been repeated attempts to try to list various tuna species for protection with the Convention on International Trade in Endangered Species of Wild Fauna and Flora (CITES, also known as the Washington Treaty).

CITES is a convention signed in 1973 among 167 countries, which regulates or prohibits trade in products made from species that are considered endangered. The convention allows for three levels of protection for species, categorized into three corresponding appendices. Appendix One species are those plants and animals that are threatened with extinction. They are offered the highest level of protection: trade in products made by these species is prohibited except in 'exceptional circumstances'.[27] Appendix Two species consist of 'species not necessarily threatened with extinction, but in which trade must be controlled in order to avoid utilization incompatible with their survival'.[28] Trade of products made from such species is permitted only with export permits issued under CITES auspices. Appendix Three contains plants and animals that are protected in at least one member country, which has asked other CITES parties for assistance in controlling the trade.

At the 1992 CITES meeting held in Kyoto, Sweden proposed the inclusion of Western Atlantic bluefin tuna as an Appendix One species and Eastern Atlantic bluefin tuna as an Appendix Two species. This was the first time that a major commercial fish stock was proposed for listing under CITES. If it were accepted, trade in Atlantic tuna would come to a halt and trade in other tuna species might also be adversely affected.

Japanese government and industry officials were outraged at the Swedish proposal, which had backing from environmental groups such as the World Wildlife Fund (now just WWF). Blaming environmentalists and foreign countries for meddling in their domestic affairs, Japanese officials mobilized protests and demonstrations in opposition to the proposal.[29] The director of

Nikkatsuren (Federation of Japan Tuna Fisheries Cooperative Associations) echoed the views of the Japanese government when he argued that the proposal was not based on scientific evidence, and regional fisheries organizations, such as the International Commission for the Conservation of Atlantic Tunas (ICCAT), were responsible for making the decisions for tuna management, not CITES.[30]

Other countries also voiced their opposition to the proposal, most importantly several members of ICCAT who agreed with the Japanese position that Atlantic tuna stock management was within their exclusive jurisdiction. Sweden finally agreed to withdraw its proposal in return for a 50 percent reduction in Atlantic tuna quotas under ICCAT for Japan, Canada and the United States.[31]

A second attempt to list Atlantic bluefin tuna in Appendix Two was made at the November 1994 meeting of CITES in Florida. The proposal was made by Kenya, reportedly at the behest of the WWF. After representations by the Japanese Fisheries Agency, Kenya agreed to withdraw the proposal and allow the issue to be settled by members of ICCAT.[32]

This controversy raised the spectre of a possible ban on bluefin tuna for the Japanese fishing industry. Senior industry and government managers have taken this incident as a warning that species of tuna – the fisheries with the greatest capture value for Japan – could be listed as endangered if Japan is not vigilant in defending its fishing interests. Of particular worry among Japanese officials are efforts to target Japan's fishing industry with moratoriums on endangered stocks, and they have resoundingly responded with a cultural defence, labelling environmentalists 'cultural imperialists' and accusing them of Japan-bashing. Given that southern bluefin tuna are listed as critically endangered and bigeye tuna as vulnerable on the Redlist of Threatened Species maintained by the World Conservation Union (IUCN), the possibility of having tuna relisted on CITES is an ever-present worry for the Japanese.[33]

Another, related issue of concern for the Japanese tuna industry is the possibility that the United States may apply its Marine Mammal Protection Act more widely to the detriment of Japan's tuna supply worldwide. The Act was designed to protect dolphins from being caught in tuna nets on the high seas and has also led to the highly publicized use of 'dolphin-friendly' tuna labelling in the United States. The Act prevents imports into the United States of tuna that have been caught with methods deemed to have a high incidence of dolphin by-catch and there is a possibility that other environmental acts, such as the Packwood-Magnuson Amendment that places trade restrictions on those states deemed to be in contravention of international environmental agreements, could be applied against Japan.[34]

Japan is wary of growing calls for protection of tuna species, as well as other marine life, whether invoked within CITES or made by the United States. In an effort to counteract this trend, Japan has a record of voting against any proposal to list a marine species for protection under CITES.[35]

For example, in the 10th Annual CITES Conference in Chile in 2002, Japan voted against all five proposals relating to protecting marine life while its proposal to have minke whales moved from Appendix One to Appendix Two, therefore permitting their trade internationally, was defeated.[36]

Termination of high seas salmon fishing in the North Pacific, 1992

The reader may remember that it was the 1952 INPFC treaty that helped revive Japanese fisheries shortly after the Pacific War and provide for the food needs of a rapidly developing country by opening up fishing grounds to the north-east of Hokkaido in the North Pacific Ocean. The subsequent 200-nm extensions by the United States and the USSR deprived the 1952 INPFC of regulatory control over halibut and herring fishing in the North Pacific since they now fell under national jurisdictions. Only the highly migratory salmon, which ventured into the mid-Pacific, remained under its mandate.

For access to fish stocks now found within national fishing zones, Japan undertook new bilateral negotiations with Canada, the Soviet Union and the United States in 1977. These nations sought to exclude Japanese fishermen from their respective zones, but the United Nations Convention on the Law of the Sea (UNCLOS) treaty, which was currently under consideration, provided Japan with limited access to excess catches beyond domestic harvesting deemed feasible under maximum sustainable yield. Although Japan was entitled to stocks deemed surplus from coastal state catch quotas, its allocation diminished considerably over the next decade.[37]

With respect to salmon, the trilateral INPFC treaty between Japan, the United States and Canada was renegotiated and a new treaty was concluded in April 1978, whereby most of the earlier provisions remained in force with the exception of a new abstention line that was demarcated at 10 degrees further east, at 175 degrees east longitude and applied only to Japanese fishing for salmon.[38] While the INPFC continued to conduct scientific surveys and monitor the abstention line, it was never vested with the powers necessary to enforce regional conservation measures, such as quotas.

Although Japanese mothership operations for salmon were still permitted under the new INPFC in the high seas (Canada, the United States and the Soviet Union fished within their own exclusive economic zones, or EEZs), some limits were placed upon their fishing grounds. Namely, Japanese fishing had to be conducted further west in the North Pacific than before. When UNCLOS was ratified in December 1982, however, a new legal precedent challenged the legitimacy of these operations altogether. Article 66 of UNCLOS declared:

> States in whose rivers anadromous stocks originate shall have the primary interest in and responsibility for such stocks ... The State of origin of anadromous stocks shall ensure their conservation by the establishment of appropriate regulatory measures for fishing ... Fisheries for

anadromous stocks shall be conducted only in waters landward of the outer limits of exclusive economic zones, except in cases where this provision would result in economic dislocation for a State other than the State of origin. With respect to such fishing beyond the outer limits of the exclusive economic zone, States concerned shall maintain consultations with a view to achieving agreement on terms and conditions of such fishing giving due regard to the conservation requirements and the needs of the State of origin in respect of these stocks.[39]

This article provided states within whose rivers salmon spawn primary control of salmon resources throughout their lifecycle, even when roaming the high seas during maturity. Except in cases of 'economic dislocation', the pelagic harvesting of salmon on the high seas was hereafter prohibited.

On the basis of Article 66, the USSR (Russia), Canada and the United States sought to renegotiate a new deal regarding the high seas harvesting of salmon. Japan argued that no new agreement was needed since the INPFC framework was already in place. Faced with the possible exclusion from salmon talks altogether, however, the Japanese delegation finally conceded to the demands of these three states of salmon origin. They agreed on 11 February 1992 to a convention that: 1 prohibited salmon fishing on the high seas; 2 restricted by-catch of salmon by fisheries targeting other species; 3 discouraged high seas salmon catch by non-parties to the new convention.[40] In 1993, the INPFC was renamed the North Pacific Anadromous Commission (NPAC), and this new body essentially prohibited Japanese salmon fishing on the high seas and allowed cooperative management of salmon resources found within national jurisdictions among the country-of-origin nations. In should be noted, however, that Japan still negotiates salmon access agreements primarily within the Russian EEZ.

Moratorium on pollack fishing in the North Pacific 'Doughnut Hole', 1994

Japan also faced restriction on its Alaskan pollack fishing in the so-called 'Doughnut Hole' in the Bering Sea in the early 1990s. The Doughnut Hole is a high seas enclave found in the North Pacific Ocean just outside the EEZ boundaries of the United States and the Soviet Union, forming a regulatory no-man's land or 'hole' (it can be seen in Figure 4.1). After being phased out of fishing grounds located within the fishing zones of both of these nations, Japanese fishermen, as well as Polish, Taiwanese, Chinese and South Korean trawlers, moved into this high seas area completely encircled by American and Russian fishing boundaries in search of pollack, cod and herring. Japanese fishermen largely targeted Alaskan pollack for *surimi* (fish paste) fish plants back in Japan, which produced the commonly eaten *kamaboko* and *oden*. This was a lucrative industry for Japanese and foreign companies supplying the Japanese market.

Fishing intensity increased within the Doughnut Hole after it was first exploited in 1985. The number of fishing boats operating on the high seas increased rapidly from twenty-five in 1982 to nearly 200 in 1990, and there were growing fears that Alaskan pollack stocks were under severe pressure and might collapse.[41]

The United States and the Soviet Union initiated talks in January 1988 to set controls on fishing effort in the high seas area. Both countries felt that many of the commercial fish stocks had originated in their own fishing zones and migrated into the Doughnut Hole, and therefore they held a stake in the management and control of these resources beyond the interests of high seas fishing states.[42] They agreed that since proper measures were taken in the EEZs, no surplus resource existed for harvesting on the high seas.[43]

The distant-water fishing nations, however, disputed coastal states stock assessment, saying that there were sufficient resources in the Doughnut Hole if properly managed. The Japanese government argued that coastal states should not be allowed to make unilateral claims to control fisheries on the high seas adjacent to their fishing zones, and that conservation and management of their resources should be based on scientific data and through discussion on an equal footing among all countries concerned, including non-coastal fishing nations.[44]

Ensuing discussions among the coastal states and distant-water fishing nations produced an agreement entitled 'The Convention on the Conservation and Management of the Pollack Resources in the Central Bering Sea', which was signed in February 1994 in Washington by the United States, Russia, Japan, Poland, the People's Republic of China (PRC), and South Korea. This convention places a moratorium on fishing in the Central Bering Sea Doughnut Hole, and sets up a management regime to govern fishing once the stocks recover.[45] The moratorium has been in place since this time.

The agreement offered coastal states considerable control over conservation and allocation decisions in the high seas adjacent to their EEZs. Conservation measures ensured that the moratorium would remain in place until the stocks were deemed to have recovered and coastal states, as an important enforcement measure, were entitled to board vessels suspected to be in violation of the agreement.[46] In order to secure Japan's agreement, however, the coastal states had to permit allocation by mutual agreement and Japanese vessels would be permitted to resume operations once the moratorium was lifted. Nonetheless, the moratorium has not been raised, which has had the net effect of depriving the Japanese of fishing grounds that had once provided an important supply of fish to the *surimi* industry.

The straddling stocks agreement, 1995

The UN's 1995 Agreement on Straddling Stocks and Highly-Migratory Stocks (also known as the Fish Stocks Agreement) also placed a number of restrictions on Japanese high seas fishing with respect to new enforcement provisions that could be applied to its vessels overseas as well as the 'precautionary principle'. As mentioned earlier, the precautionary principle stated that where threats of serious or irreversible damage to fish stocks existed, lack of full scientific certainty could not be used as a reason for postponing measures to prevent environmental degradation. This would make it easier for coastal states or international fisheries management organizations to place moratoriums on commercial fishing when there is a threat of overfishing.

The UN elected to host a conference on straddling stocks in April 1993 as a direct result of a declaration of the Rio Summit that called for a new agreement to regulate high seas fishing as well as ongoing disputes over the management of fish stocks that straddle the boundaries between EEZs and the high seas, such as pollack, tuna and salmon. Governmental and NGO delegates from over one hundred nations participated in the negotiations, which lasted a total of six sessions over two years.[47]

The conference was generally divided into two camps: coastal states and distant-water fishing countries. Coastal states, led by Canada and countries in Latin America and the Pacific Islands, desired a legally binding convention that would subject the freedom of high seas fishing to the interest of coastal states, arguing that they were charged with preserving and managing fisheries resources within their EEZs under UNCLOS.[48] On the other hand, distant-water fishing supporters – including Japan, PRC, South Korea and the European Union – called for 'soft' regulations on the high seas as well as an agreement that would form a set of non-binding guidelines only.[49]

The Japanese delegation aimed to minimize restrictions regulating the high seas. The Fisheries Agency strongly opposed the extension of coastal state responsibilities and rights beyond 200-nm EEZs into the high seas, including the right to board vessels belonging to other nations on the high seas.[50] The Japanese government also teamed up with European delegates to oppose the precautionary approach being considered at the conference, stating that it could lead to the adoption of greater moratoriums for high seas fisheries.[51] Instead, Japan advocated the creation of regional fisheries organizations that would have the right to determine appropriate measures for the conservation of stocks given the particular situation of each ocean region.[52] They also proposed stronger international regulations to regulate the re-flagging of fishing vessels to evade controls established by national or regional agencies.[53]

The final agreement involved a number of compromises on both sides of the debate, but largely adopted the provisions sought by coastal states. In accord with Japan's wishes, Article 7 of the UN Fish Stocks Agreement called for the establishment of conservation and management regimes on the high seas that must be based on cooperative agreement between fishing nations in the area and coastal states in the region.[54] The agreement stated that only nations that are members of such regional management organizations would be permitted access to the resources under the organization's control. The treaty also permitted on-board inspections of vessels on the high seas, even if the vessel is not owned by the inspecting state, in order to ensure compliance with regional conservation agreements.[55] This was considered to be a major concession made by the Japanese delegation at the talks.[56]

Finally, the very first codification of the precautionary approach was included in an international fisheries agreement. Article 6 declared that it would provide the guiding basis of future high seas fisheries management:

> States shall be more cautious when information is uncertain, unreliable or inadequate. *The absence of adequate scientific information shall not be used as a reason for postponing or failing to take conservation and management measures* ... If a natural phenomenon has a significant adverse impact on the status of straddling fish stocks or highly migratory fish stocks, States shall adopt conservation and management measures on an emergency basis to ensure that fishing activity does not exacerbate such adverse impact. States shall also adopt such measures on an emergency basis where fishing activity presents a serious threat to the sustainability of such stocks.[57]

This Article paved the way for fisheries restrictions, including moratoriums, to be applied when a fisheries resource is under significant threat of depletion in high seas waters. It also stated that the lack of scientific evidence could not be used as an excuse for not adopting conservation measures when a resource is suffering an 'adverse impact' due to natural or man-made causes such as fishing. Given that Tokyo had repeatedly argued in the case of whaling and driftnets that there was insufficient scientific evidence to support a ban on these activities, this article has particular significance for Japan. The Japanese can no longer rely upon this rationale to stave off future restrictions on their fishing activities on the high seas.

Conclusion

Changing international values with respect to ocean resources in conjunction with increasingly restrictive measures adopted in fisheries conventions posed a dual challenge to the Japanese government. How might national policy on food security and fisheries self-sufficiency be better articulated and defended? What kinds of countermeasures might be adopted to protect its fishing interests abroad in the wake of international restrictions? These will be dealt with in turn in the next chapter.

Notes

1 Jonathan B. Wiener, 'Precaution', *Oxford Handbook of International Law*, Daniel Bodansky, ed. Oxford: Oxford University Press, 2007, 598.
2 Ibid., 604–6.
3 Ibid., 599.
4 See Harry Scheiber, *Inter-Allied Conflicts and International Law, 1945–53: The Occupation Command's Revival of Japanese Whaling and Marine Fisheries.*

Taipei: Academia Sinica, 2001, for a detailed study of the resumption of whaling under the Occupation authorities.

5 Masayuki Komatsu and Shigeko Misaki, *Whales and the Japanese*. Tokyo: Japan Whaling Association, 2003, 66; and Arne Kalland and Brian Moeran, *Japanese Whaling: End of an Era?* London: Curzon Press, 1992, 89–90.

6 International Convention for the Regulation of Whaling, 'Preamble', Washington, 2 December 1946, www.iwcoffice.org/Convention.htm. It manages all species of baleen and toothed whales with the exception of small cetaceans, such as dolphins and porpoises.

7 The BWU was a measurement of the amount of oil one blue whale could yield: equivalent to two fin whales, two-and-a-half humpback whales and six sei whales. Japan Whaling Association (JWA), www.whaling.jp/english/history.html#04.

8 Steinar Andresen, 'Science and Politics in the International Management of Whales', *Marine Policy* vol. 13 (April 1989): 104.

9 Sidney Holt, 'Whale Mining, Whale Saving', *Marine Policy* vol. 9 (July 1985): 12.

10 William Aron, William Burke and Milton Freeman, 'The Whaling Issue', *Marine Policy* vol. 24 (May 2000): 180.

11 Aron *et al.*, 'The Whaling Issue', 180.

12 Ibid., 180.

13 The Pelly Amendment (1973) allows for an embargo on all fish and wildlife products from any country pursuing policies that weaken international conservation measures, whereas the Packwood-Magnuson Amendment (1978) requires that once the secretary of commerce determines that a country has diminished the effectiveness of an international conservation convention, it must lose at least 50 percent of its fishing quota in the US exclusive economic zone.

14 Komatsu and Misaki, *Whales and the Japanese*, 90.

15 Anthony Bergin and Marcus Haward, *Japan's Tuna Fishing Industry: A Setting Sun or a New Dawn?* New York: Nova Science Publishers, 1996, 96.

16 A detailed background of the Greenpeace campaign can be found at 'Campaigns', www.greenpeacefoundation.com/action/.

17 Lynda Paul, *The Impact of Driftnet Fishing on Sustainable Fisheries: High Seas Driftnetting, the Plunder of the Global Commons*. Kailua, HI: Earthtrust, 1994, 5.

18 It may be mentioned that this law was passed in spite of the INPFC's scientific mandate to permit observers on Japanese vessels, which suggests that the United States invested little faith in the INPFC's monitoring or conservation capacity.

19 United Nations, UN Doc. A/RES/44/225, 85th plenary meeting, 22 December 1989, www.un.org/documents/ga/res/44/a44r225.htm.

20 Moritaka Hayashi, 'Fisheries in the North Pacific: Japan at a Turning Point', *Ocean Development and International Law* vol. 22 (1991): 358–59.

21 United Nations, UN Doc. A/C.2/44/L.30/Rev.1 (1989), found in ibid., 359.

22 'Gillnet Fishing Treaty', undated, gurukul.ucc.american.edu/TED/GILLNET.HTM.

23 Some environmentalists fear that readjustments might mean the sale of Japanese driftnetting boats and equipment to other countries which do not comply with the UN Resolution, such as Taiwan or the PRC. They also report that some boats now operate under the flags of other nations although the owners remain Japanese. Although the ban on driftnets was a devastating blow to Japanese squid and salmon trawlers in the North Pacific, other vessels from the PRC, Taiwan and South Korea continue to use driftnets illegally, catching valuable salmon and other species to fill the gap left in the Japanese market. In some ways the problem has not changed, only the perpetrators.

24 William Burke, Milton Freeberg and Edward Miles, 'United Nations Resolutions on Driftnet Fishing: An Unsustainable Precedent for High Seas and Coastal Fisheries Management', *Ocean Development and International Law* vol. 25, no. 127 (1994): 127–86.

25 This was widely confirmed at interviews with officials at both the Fisheries Agency and the Japan Fisheries Association, who raised this issue and its role as precedent in connection to the ongoing whaling dispute.
26 See for example, Nagasaki Fukuzō, 'Fisheries and Environmentalism', Institute for Cetacean Research (ICR), circular, written by the Director of ICR, 1994.
27 Convention on International Trade of Endangered Species, www.cites.org/eng/disc/how.shtml.
28 Ibid.
29 *Aera*, '"Toro kūna" no kyokyo jitsujitsu – suĕden no chōsenjō suisan shigen' (Do Not Eat Tuna!' – The Shrewd Strategy of Sweden on Fisheries Resources), 17 March 1992, 6.
30 *Asahi Shimbun*, 'Sushi ga maguro wo zetsumetsu saseru?! kyūbujō, kuromaguro gyokisei' (Is Sushi Making Tuna Extinct?! The Rapidly Emerging Regulation on Bluefin Tuna), 20 January 1992.
31 *Yomiuri News Service*, '4 CITES Members Agree to 50 Percent Cut in Bluefin Tuna Catches', 10 March 1992.
32 Sandra Tarte, *Japan's Aid Diplomacy and the Pacific Islands*. Canberra: National Center for Development Studies, Australian National University, 1998, 137.
33 The World Conservation Union (IUCN) Redlist (www.redlist.org) has provided taxonomic, conservation status and distribution information on plants and animals for the last forty years. It is designed to determine the relative risk of extinction and its primary purpose is to catalogue and highlight those species that are facing a higher risk of global extinction. IUCN is a global alliance of governmental, academic and NGO members, including the United Nations Environment Programme.
34 Bergin and Haward, *Japan's Tuna Fishing Industry*, 91.
35 BBC, 'Politics Threaten Conservation Talks', 10 November 1992.
36 Fisheries Agency, Press Release, 14 November 2002.
37 Aoki Hisano, '200 kairi to nihon no gyogyō: beikoku oyobi soren no tainichi gyotaku wariate o chūshin ni shite' (200 Mile EEZs and the Japanese Fishing Industry: A Focus on Quota Allocations with Respect to the United States and the Soviet Union), *Kokusai Kankei Kenkyū* (International Relations Research) no. 7 (1984): 31–56.
38 International North Pacific Fisheries Commission, *International North Pacific Fisheries Commission Handbook*. Vancouver: International North Pacific Fisheries Commission, 8–11.
39 United Nations Convention on the Law of the Sea, 'Agreement Relating to the Implementation of Part XI of the Convention', Article 56, www.un.org/Depts/los/convention_agreements/texts/unclos/closindx.htm.
40 Hayashi, 'Fisheries in the North Pacific', 352.
41 Ibid., 351–52.
42 Federal News Service, 'Statement of David Benton, Director Office of External and International Fisheries, Alaska Department of Fish and Game, Before the Senate Committee on Commerce, Science, and Transportation', 21 April 1994.
43 Hayashi, 'Fisheries in the North Pacific', 355.
44 Japanese Commissioner's Address to the Opening Plenary Session on the 37th Annual Meeting of the International North Pacific Fisheries Commission, Vancouver, 6 November 1990, 3, found in ibid., 352.
45 Statement of David Benton, 1994.
46 Bergin and Haward, *Japan's Tuna Fishing Industry*, 98–99.
47 United Nations, 'Conference on Straddling Stocks and Highly-Migratory Stocks', www.un.org/Depts/los/.
48 Inter Press Service, 'U.N. Talks to Tackle Problem of Overfishing', 9 July 1993.

49 *Asahi Shimbun*, 'Kōkai gyogyō wo kokusai jōyaku de kisei: raiaki no seitei mezasu, kokurenkaigi ga hōshin kettei' (Regulation of High-seas Fisheries by an International Treaty: Aiming at Enactment by Next Fall, a United Nations Conference Decides on Principles), 28 August 1994.
50 *Asahi Shimbun*, 'Izondo takamaru nihyaku kairi-nai gyogyō: shigen kanri kyōtei saitaku de nihon' (Increasing Dependence on Fishing within 200nm EEZs: Japan after the Adoption of an Agreement on Resource Management), 5 August 2005.
51 Inter Press Service, 'Progress Slow at U.N. Fisheries Conference', 16 July 1993.
52 *Asahi Shimbun*, 'Gyogyō kisei nado saki okuri: kōkai no shigen de tairitsu, kokuren kaigi heimaku' (Postponement on Fisheries Regulations: A United Nations Conference Ends after Conflict on Marine Resources), 31 July 1993.
53 *United Press International*, 'Japan Calls for Global Control of High-Seas Fishing', 29 March 1994.
54 United Nations, *Agreement for the Implementation of the Provisions of the United Nations Convention on the Law of the Sea of 10 December 1982 Relating to the Conservation and Management of Straddling Fish Stocks and Highly Migratory Fish Stocks*, A/CONF.164/37.
55 Ibid., Article 21.
56 Japan Economic Newswire, 'Japan Conditionally Backs U.N. Fishing Violation Curb', 25 July 1995.
57 United Nations, *The Conservation and Management of Straddling Fish Stocks and Highly Migratory Fish Stocks*, Article 6, emphasis added.

Bibliography

Andresen, Steinar. 'Science and Politics in the International Management of Whales', *Marine Policy* vol. 13 (April 1989): 99–117.
Bergin, Anthony and Marcus Haward. *Japan's Tuna Fishing Industry: A Setting Sun or a New Dawn?* New York: Nova Science Publishers, 1996.
Burke, William, Milton Freeberg and Edward Miles. 'United Nations Resolutions on Driftnet Fishing: An Unsustainable Precedent for High Seas and Coastal Fisheries Management', *Ocean Development and International Law* vol. 25, no. 127 (1994): 127–86.
Kalland, Arne and Brian Moerān. *Japanese Whaling: End of an Era?* London: Curzon Press, 1992.
Komatsu, Masayuki and Shigeko Misaki. *Whales and the Japanese.* Tokyo: Japan Whaling Association, 2003.
Paul, Lynda. *The Impact of Driftnet Fishing on Sustainable Fisheries: High Seas Driftnetting, the Plunder of the Global Commons.* Kailua, HI: Earthtrust, 1994.
Scheiber, Harry. *Inter-Allied Conflicts and International Law, 1945–53: The Occupation Command's Revival of Japanese Whaling and Marine Fisheries.* Taipei: Academia Sinica, 2001.
Tarte, Sandra. *Japan's Aid Diplomacy and the Pacific Islands.* Canberra: National Center for Development Studies, Australian National University, 1998.

6 Comprehensive security as national policy and Japan's new fisheries strategy

After the widespread declaration of exclusive economic zones (EEZs) and the imposition of high seas fisheries controls and moratoriums, Japanese policy makers were once again confronted with the problem of how to guarantee a stable supply of fisheries for Japanese consumption. Unlike the political situation in the immediate post-war period, however, Japan did not have the influential backing of the United States when making its claims internationally. Japan was alone in its pursuit of a new fisheries strategy in the post-EEZ era.

The motives behind the earlier fisheries strategy were re-evaluated in the 1970s in light of a changed domestic and international environment. Japan was becoming more dependent upon imports for meeting domestic demand while the application of greater fishing restrictions abroad served to undermine Japan's quest for self-sufficiency. A new comprehensive security strategy was adopted, which reaffirmed food security and self-sufficiency as one of the cornerstones of national policy. Although the motives for international fishing policy remained focused on self-sufficiency, new means were adopted to realize this goal.

Since Japan lost many valuable distant-water fishing grounds with the onset of the enclosure movement as well as free access to fisheries that were now under restrictive protection, it had to look for new fishing grounds and apply an alternative approach for finding sources of fish. This new fishing strategy can be divided into four main parts: developing coastal fisheries more intensively, managing increasing imports, negotiating new bilateral agreements and promoting open access to fisheries at multilateral forums. Each of these strategies will be examined in greater detail later in the chapter and in Chapter 7.

Self-sufficiency, food control and comprehensive security policy

The idea of self-sufficiency in food production attracted widespread attention in Japan after a series of crises in the 1970s highlighted Japan's vulnerability with respect to imported resources. Not only did the increasing application of restrictions on distant-water fishing create a great deal of anxiety within the Japanese fishing industry during this time, but in 1972 an international food crisis shocked the world when heavy demand, created by a worldwide spate of crop failures and huge grain sales to the Soviet Union and China, depleted world food reserves.[1] Consecutive oil crises in 1973 and 1979 also added higher costs to food production, which had become quite dependent upon petroleum products for fuel and fertilizer, while the US soybean embargo in 1973 came as a great surprise to Japan.[2] Given that Japan was almost completely reliant upon the United States for soybean imports, the soybean crisis shattered any complacency that adequate purchasing power alone could guarantee security of supply. Thus these crises served to remind Japanese policy makers of the unreliability of foreign resource supplies and the benefits of maintaining self-reliance with respect to food.

Such crises prompted a rethinking of resource and food security in terms of 'comprehensive national security'. At a time when many Japanese started to question the global leadership role of the United States after its deteriorating trade performance and failing military venture in Vietnam, Prime Minister Ōhira commissioned a number of studies in the late 1970s that investigated possible scenarios in which Japan could exercise greater influence internationally and take a more proactive role in defending its interests abroad. The most influential of these reports, *The Report on Comprehensive National Security*, was released by the Inoki Task Force in July 1980 and was subsequently endorsed by Ōhira's successor, Prime Minister Suzuki Zenkō.[3]

'Comprehensive security' differs from the conventional use of the term 'security', in that it places greater emphasis on the economic dimensions of national stability than on military ones. The Inoki Task Force used the term to examine the ways economic instruments may be used to promote stability as well as the means that a stable economic setting leads to greater security for all. Their report suggested that comprehensive security supported the role of foreign policy in securing access to resources, safe trading lanes, stable trading partners and fostering future industries as essential elements for a safe and stable world.[4]

Comprehensive security advocated three levels of security effort: self-reliant efforts; efforts to turn the international environment into a favourable one through the creation of a more peaceful international order; and intermediary measures that fostered cooperation with countries important to Japan and sharing common interests.[5] While not ignoring the central position occupied by the Japan-US Security Treaty in its security interests, Japan nonetheless started to reappraise the basis of its foreign policy through a consideration of economic determinants of national stability.

The Inoki Task Force also examined the role of food security in Japan's foreign policy agenda, raising this issue to a level of national security and state priority.[6] Food security, however, was not defined simply in terms of self-sufficiency as had been done for previous decades. Although the report concludes that some degree of self-reliance is desirable in Japan's food supply, 'it would be too hasty a conclusion to insist that Japan must increase its level of self-sufficiency'.[7] Instead, the Inoki report suggested that the government should develop contingency plans for possible crises, such as the paralysis of maritime transportation, a world food crisis caused by poor harvests or excessive demand, worsening diplomatic relations leading to an interruption in trade, or a Malthusian crisis when world population growth outstrips the rise in food production.[8]

Thus, the report recommended that Japan should aim to keep potential production capacity at the highest possible level in the event of a national emergency rather than concentrate on the degree of self-sufficiency that Japan enjoys during normal times.[9] The Inoki Task Force departed from conventional wisdom somewhat by recommending that food security policy should strike a balance between supporting the domestic production of agriculture and fisheries while promoting the principle of free trade to meet overall national demand.[10] This governmental commission harkened back to many of the same issues regarding self-sufficiency that were raised in Meiji-era debates as reviewed in Chapter 1.

Despite this reference to allowing managed trade to help guarantee national food security, the political elite in Nagatachō opted to play down these recommendations and decided instead to emphasize the Inoki Report's findings on self-sufficiency. On 8 April 1980, the House of Representatives unanimously adopted a resolution urging the government to 'take all appropriate measures to develop agriculture and fisheries, and to promote self-sufficiency in food as an important element of the people's security'.[11] Policy makers were not yet ready to abandon the goal of self-sufficiency planning.

The Inoki Report's recommendations on fostering reliable trade links were perhaps a pragmatic recognition that Japan's food import situation had changed quite significantly in the preceding decade. In spite of various economic reforms following Japan's defeat in the Pacific War, the Japanese economy remained essentially autarkic in nature. In 1960, however, the Japanese government embarked upon a programme to liberalize trade in selected areas of the food market, with the important exception of rice. In 1964, Japan adopted Article 8 status in the International Monetary Fund and Article 11 status in the General Agreement on Tariffs and Trade (GATT), which required that a nation impose neither exchange restrictions nor import restraints in its trade relations.[12] These actions marked the beginning of the Japanese economy's journey toward a more liberal domestic market for agricultural and fisheries trade. As a result, in spite of the official rhetoric on self-sufficiency, Japan began to import a greater amount of foodstuffs to the extent that Japan's self-sufficiency rates, measured in terms of calories, fell from 80 percent in 1955 to 51 percent in 1972.[13] The most significant increases in imports were for cereals, soybeans and sugar, whereas fisheries imports only had a modest rise in calorific terms during this period.[14]

In view of rising food imports in the 1960s and 1970s, why did Japan adhere to a policy of supporting self-sufficiency? Did the government feel that it could somehow rectify the balance of trade and domestic production in favour of a higher degree of self-sufficiency, or did it merely seek to support domestic food production to the greatest extent possible? To answer these questions, we must examine the role of path dependency in the food control system, which created strong domestic incentives for the continuation of the policy of self-sufficiency even in the face of increasing trade dependency.

Food self-sufficiency continued as a primary aim of MAFF's food control system for several decades after the war despite the elimination of any risk of food shortages on the scale witnessed during the Occupation, and despite the changing tastes of consumers who were becoming increasingly urbanized and more wealthy. Although daily per capita food supply fell below 2,000 calories in the immediate post-war period, price controls on rice and the expansion of cultivatable land increased domestic production to pre-war levels by 1955 whereby Japan achieved about 80 percent self-sufficiency.[15] Rice attained 100 percent self-sufficiency by 1955 (12 million tonnes of production) largely through various incentives given to production through the food control system, but this was achieved at a cost whereby consumers paid three times the world price.[16] Thus the agricultural and fisheries industries had reformed to the extent that, by 1952, most food needs could be met through domestic production and neither famine nor malnourishment threatened Japan any longer.

By the 1970s, Japanese dietary habits changed enough that some observers began to question the very basis of the policy of self-sufficiency. Food consumption habits underwent drastic changes during the 1955–75 period: consumption of cereals, potatoes and soybeans declined considerably, while concurrently the consumption of meat, eggs, milk, dairy products, and oils and fats increased.[17] In addition, rice consumption fell drastically over the same period. Whereas annual per capita intake in 1955 was 111 kg, by 1975 this had dropped to 88 kg.[18] Fish consumption, on the other hand, rose in the diet from 26 kg per person in 1955 to 35 kg per person in 1975.[19] Not only was rice declining in importance as a staple crop, but the agricultural industry was losing its comparative advantage relative to other industrial countries due to its focus on capital-intensive rice production.[20] In other words, the price controls and tax benefits on rice production offered under the food control system were creating costly inefficiencies in the economy for the sake of self-sufficiency.

Despite these changes in the Japanese diet over several decades and the effect this could have had upon planning for self-sufficiency, the food control system, which supported rice production in particular, continued to receive support from the Japanese government even when such economic inefficiencies were becoming more evident. The preservation of the policy of self-sufficiency over many decades is largely attributable to the vested interest that politically connected and influential agricultural cooperatives had in the continuance of the food control system. The nationwide Organization of Agricultural Cooperative Unions (Nōgyō kyōdō kumiai, or Nōkyō for short) has had an entrenched interest in the maintenance of the food control system, especially since it was the primary beneficiary of government subsidies and price controls for rice collection and marketing that were authorized by the Food Control Law. 'The preservation of these vested rights has been consistently behind [Nōkyō's] approach to the whole question of FC [food control] reform. In essence, the FC system and Nōkyō supported each other over many decades'.[21]

Nōkyō has used political connections to maintain the system of food control in a relationship that can be deemed 'path dependent'. Nōkyō has strong connections with the ruling Liberal Democratic Party, often successfully backing politicians who were sympathetic to the cooperative's interests.[22] These politicians have exerted influence on not just the legislative process concerning food control, but also on budgetary allocations and the policies of administrative agencies, most importantly the Ministry of Agriculture, Forestry and Fisheries (MAFF). This situation appears to be a classic case of path dependence whereby a particular course of action is very difficult to change over time due, in part, to the status quo bias built into the political institutions.[23] Thus, Nōkyō has played a large role in supporting the system of food control throughout the post-war period and the premise of self-sufficiency upon which it is based.

Thus, national policy on self-sufficiency has continued as a result of sectoral interests within the industries dependent upon the food control system, most importantly rice. These interests and dependencies were created, however, as a result of the self-sufficiency policy adopted in the early post-war period during a time when food shortages were of much greater concern. Despite the sectoral basis for continued support for the food control system, self-sufficiency can be viewed as a *national* interest given that the politicians and bureaucrats continued to champion this cause as a component of the broader foreign policy goal of comprehensive security.

The goal of food security has not, however, been an attempt necessarily to increase the rate of self-sufficiency, but to keep Japan as self-sufficient as possible given a limited resource base. As an official at MOFA described this reasoning:

> The Japanese have a common belief that we need to make the best use of available resources. Therefore, any action that may discourage people's motivation to utilize these resources will face some sort of difficulties. The policy of self sufficiency does not mean that Japan totally safeguards its producers irrespectively, but that self sufficiency should not be allowed to fall below a minimum level.[24]

In other words, Japan aims to make use of all resources – fisheries or otherwise – that are available to the nation and aims to ensure that access to these resources is not reduced or hampered. Self-sufficiency has thus become a goal to ensure a minimum level of domestic production to meet national food requirements and not as a policy necessarily to increase that ratio.

With self-sufficiency firmly entrenched in the thinking of Japanese fisheries officials, a new fisheries strategy was adopted in subsequent fisheries negotiations and policy making to achieve this national goal. Japanese officials devised a four-part plan in response to various fisheries restrictions, which aimed to develop coastal fisheries more intensively, manage the amount of international imports, establish new bilateral agreements and promote the notion of increased or free access to fisheries resources at multilateral forums and organizations. The domestic components of this plan will be investigated henceforth, whereas the international components involving diplomacy, negotiation and new international treaties will be examined in turn in Chapter 7.

Domestic food security policy in fisheries

Of the aforementioned four-point plan with respect to resource security, the expansion of coastal fisheries was the easiest of these strategies to achieve. The promulgation of a new fisheries law in 1977 provided Japan with a much larger fisheries zone for domestic fisheries to exploit. The import of fish was also possible due to expanded fishing capacity internationally resulting from the enclosure movement, although at greater cost than before.

1 Expanding the domestic fisheries effort

Domestic fishing grounds grew substantially with the establishment of Japan's new exclusive economic zone. The 1977 Provisional Fisheries Law expanded Japanese fisheries jurisdiction to 376,000 square km of ocean space, or fifty times as large an area as it controlled under the 3-mile territorial sea rule.[25] This provided Japan with the world's seventh largest fisheries zone.

Domestic fisheries, including coastal, offshore and aquaculture production, increased their effort over the years after 1977 and far surpassed the distant-water catches that were steadily declining in importance. This expanded effort, however, peaked in the late 1980s and has been in decline since. According to Figure 6.1, domestic fisheries produced 7,866,000 metric tons in 1977 as opposed to 2,683,000 metric tons for distant-water fisheries. By 2000, however, the disparity in catches had grown even wider, with domestic and distant-water catches totalling, respectively, 5,398,000 metric tons and 855,000 metric tons. Whereas domestic fisheries effort expanded in the 1980s, in part due to large catches of sardines during these years, catch totals peaked in 1988 and have been steadily declining since then. Meanwhile, distant-water catches have been in steady decline since 1977, although the rate of that decline has not been as precipitous as that for domestic production.

The Japanese government has offered significant financial and political support in an effort to boost domestic production. In 2000, the government made 314 billion yen (US$2.93 billion) in transfer payments to the domestic fishing industry.[26] This is the world's highest total for government subsidies to fisheries and does not include other allowances that may be offered, such as tax breaks, import/export controls and community enhancement programmes. The ruling Liberal Democratic Party has also targeted rural communities, including fishing villages and towns, in its campaign for community development that will ensure future commitment to projects aimed at infrastructure and lifestyle improvement, in an attempt to stem the tide of people leaving fishing and other rural communities to the cities in search of new careers.[27]

Another programme the Japanese government introduced to assist domestic fishers in the mid-1970s onward was a vessel reduction and buy-back scheme. This programme amounted to a range of subsidies and financing aimed at making domestic fishermen more competitive in an increasingly global marketplace while concurrently reducing fishing capacity that was widely recognized as being unsustainably high in terms of both economic scales of production and biological limits on catch capacity. An examination of the buy-back and financing programmes for the distant-water tuna fleet will show that these goals were rarely achieved in full, resulting in mixed effects not always beneficial to domestic producers.

Facing rising labour, fuel and access fee costs in the mid-1970s and competition from imports from South Korea and Taiwan in the 1980s, Japanese tuna fishers, both distant-water and domestic, struggled to make their

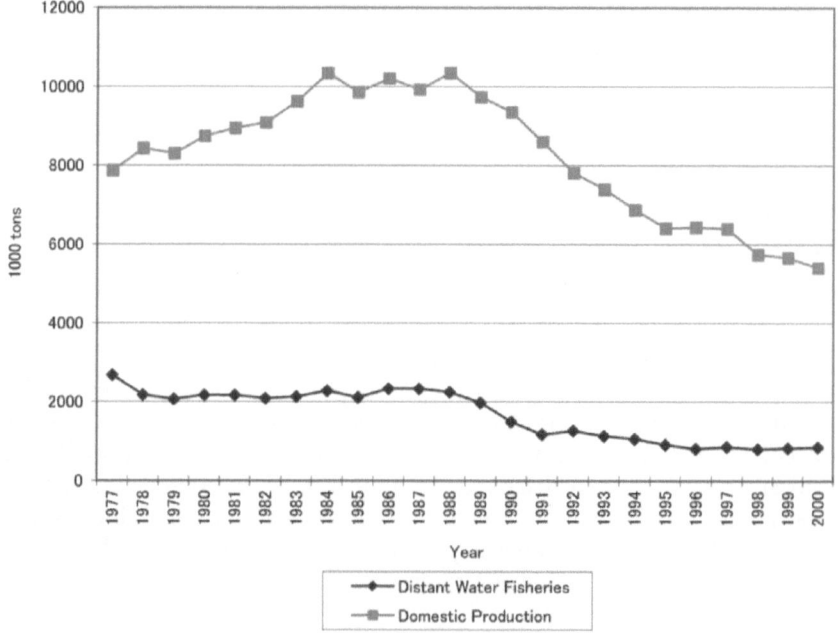

Figure 6.1 Domestic and distant-water catches, 1977–2000
Source: Ministry of Agriculture, Forestry and Fisheries, *Gyogyō yōshokugyō seisan tōkei nenpō* (Statistical Yearbook of Fisheries and Aquaculture Production). Tokyo: Nōrinsuisanshō tōkei jōhōbu, 1977–2000)

operations profitable. Japanese officials sought to improve the fishers' plight by reducing the size of the distant-water tuna fleet to make it more cost efficient and avoid creating an oversupply of tuna in the Japanese market that would be disadvantageous to domestic fishers. The Fisheries Agency adopted a two-prong approach to achieve this: fewer licences were granted to distant-water tuna vessels and a low-interest loan programme was established to compensate owners and crew members of vessels that were retired.[28]

This approach led to several unforeseen results. Many owners who scrapped their vessels purchased entirely new ones that were more fuel efficient and required fewer crew members to operate. These purchases were financed by the government's buy-back programme as well as another government loan programme for ship construction which appeared to work at cross-purposes with the vessel retirement scheme. In all, the Japanese distant-water tuna fleet declined by less than 10 percent – a mere 100 vessels – between 1985 and 1992.[29]

Other owners opted not to scrap their vessels but to export them to rival fleets in South Korea, Taiwan and flag-of-convenience states. This has resulted in a transfer, not reduction, of fishing capacity away from Japan to countries that export tuna to Japan. Two hundred reflagged vessels, registered in Panama and Honduras and managed by companies in Japan, South Korea and Taiwan, have been fishing for tuna in the Pacific, which accounted for about 22 percent of the frozen sashimi tuna imported into Japan in 1991.[30] Thus, despite the Fisheries Agency's efforts, the buy-back programme had effectively led to a shift in some Japanese fishing capacity to rival fishing fleets as well as only a minimal decline in fishing capacity that has not eliminated the problem of oversupply for domestic producers.

2 Managing marine product imports

Japan also turned to managing importing valuable fisheries products as a way to counterbalance the loss of overseas fisheries. Table 6.1 illustrates the shift in fisheries supply away from domestic and distant-water production in the 1960s and 1970s to a greater reliance upon imported fish products by the late 1980s. Whereas domestic production supplied over 96 percent of the fish products consumed in 1965, imports amounted to a meagre 5 percent.[31] After the widespread adoption of EEZs in the mid- to late 1970s, however, imports increased substantially, constituting a total of 56 percent of all fish consumed in Japan in 1989, whereas domestic production fell to 34 percent. This represents a significant structural change over two decades in favour of imports supplying the Japanese market.

An examination of the trends for fisheries imports also reveals that the costs for fisheries products rose in proportion to the volume of these imports. Whereas Japan imported a total value of US$232 million in fisheries products in 1977, this jumped to a total of US$2.4 billion by 1999, as seen in Figure 6.2. The rate of increase for the value of imports is similar to the volume,

Table 6.1 Demand and supply of fisheries products in Japan, 1965–89

Year	Demand		Supply			Domestic consumption ratios		
	Consumption	Export	Domestic Production	Fish farming	Imports	Ratio of domestic production (%)	Ratio of fish farming (%)	Ratio of imports (%)
1965	1,973	260	1,904	226	103	96.5	11.5	5.2
1969	2,114	162	1,803	279	194	85.3	13.2	9.2
1973	2,661	341	2,117	359	526	79.6	13.5	19.8
1977	2,931	196	2,179	429	818	74.3	14.6	27.9
1981	2,928	158	1,616	478	998	55.2	16.3	34.1
1985	3,329	151	1,547	566	1,356	46.5	17.0	40.7
1989	3,352	345	1,127	686	1,870	33.6	20.5	55.8

Note: Unit 1,000 tonnes.
Source: Hasegawa Akira, 'Kokusaika-jidai no nihon-suisangyō no dōkō to kadai: yūnyū mondai wo chūshin ni shite' (Trends and Issues for Japanese Fisheries in an Age of Internationalization: Focus on the Question of Imports), in Yasuhiko Taki (ed.) *Sekai no naka no nihon gyogyō* (Japanese Fisheries in a Global Perspective). Tokyo: Seisandō, 1993, 2)

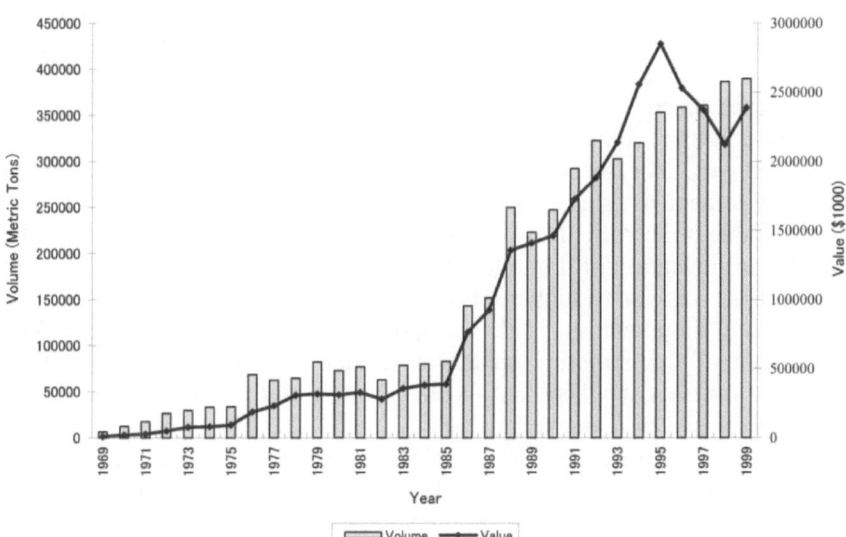

Figure 6.2 Japanese fisheries imports (value and volume), 1969–99
Source: FAO Fishstat Database, www.fao.org/fi/statist/FISOFT/FISHPLUS.asp

suggesting that the price of fish did not increase substantially during this time period, with the exception of the period between 1993 and 1995.[32] The lion's share of the cost of these imports has been spent on four products: shrimp, salmon, crab and tuna.

Rising fisheries imports into Japan were made possible due to a number of factors. Most important of these was the rise in value of the yen in the 1970s and 1980s that made the import of fish products relatively cheaper than in previous years. Other factors included the so-called 'structuralization' of fisheries imports in the 1980s, the gradual liberalization of trade in fisheries products at successive rounds of the GATT, and the rationalization of the distant-water fishing industry within Japan.

The rising value of the yen in the 1970s and 1980s created an impetus toward the greater importation of fisheries products from overseas. The yen continued to rise between 1971 and 1995, from the peg it was set at 360 yen to the dollar in 1949, to 308 yen in 1971 and 210 yen in 1978, finally reaching past 150 yen in 1987 after the Plaza Accord Agreement in 1985 revalued a stronger yen against a weakened dollar.[33] The rise in the value of the yen entailed a concomitant fall in the price of seafood imports. For example, the index price for shrimp, which constituted nearly 30 percent of the value of seafood imports in 1985, fell by 25 percent over the 1984–87 period, whereas the volume increased by 45 percent.[34] The Japanese market for seafood expanded with this new-found purchasing power.

Another factor facilitating the rise in imports was the structuralization of the fisheries import system in the 1980s. Horiguchi Kenji describes this process of market structuring as involving three steps. The first step was the gradual development of fisheries capacity, fish farms and production facilities in foreign countries aimed at servicing a Japanese market empowered by a high-valued yen. The second step involved the increase of Japanese direct foreign investment in fishing plants, facilities and businesses abroad that created a reliable supply chain for seafood products.[35] The final step was the gradual transformation of domestic retailing of fish products that relied upon cheap seafood imports of high quality, particularly as supermarkets began to replace small shop owners as the primary retailers of fish produce.[36] The development of this structure created a reliable link between the Japanese market and overseas suppliers that facilitated trade as the demand for fisheries products increased over the years.

The lowering of trade barriers worldwide also provided an impetus for greater trade in fish products. Since 1958, GATT trade liberalization measures have been agreed upon for most major fisheries products including the most important produce for Japan: shrimp, salmon, tuna and crab.[37] It is curious to note, however, that Japan still maintains tariffs for many fish products and the United States has threatened to raise this issue before the World Trade Organization on the grounds that it is a discriminatory trade practice.[38] The Japanese government uses such tariffs in an effort to protect domestic fishermen from competition from cheap imports in similar products, although the rates are relatively small, if even applied at all.[39] Whereas Japanese government officials are rather sceptical of the possibility of actually raising the rate of self-sufficiency, they still aim to minimize the fall in this rate by protecting domestic production to the greatest extent possible through tariff protection and similar measures.[40]

Another factor that facilitated the market shift toward fisheries imports was the effect of a rationalization process undertaken with the distant-water fishing industry itself. These fishing companies not only faced rising oil costs during both oil crises in the 1970s, which had substantial impacts upon the costs of operations, but the wages of fishermen on long distance operations – which accounted for roughly a third of costs for trawlers – were rising in proportion to the increasing standard of living within Japan's cities.[41] By 1991, the wage level for distant-water fishermen was roughly equal to those of other industries in Japan, whereas it was 40 percent lower in 1970.[42] Facing rising costs for international operations, it became much cheaper for distant-water fishing companies to import marine products from abroad than to make fishing runs to foreign fishing grounds.

Japan's two largest high seas fishing companies, Taiyō Gyogyō and Nippon Suisan, adopted new corporate strategies to cope with rising costs and shrinking fishing grounds. Taiyō continued to diversify its operations into other agricultural food sectors, real estate, and even a baseball team.[43] Meanwhile, the company has sustained its fishing interests through the establishment of an extensive international import operation and more intensive fish farming. By 1985, Taiyō was among the foremost harvesters of cultivated fish in the world, producing 10 percent of the company's annual catch through fish farms at various locations in Japanese waters.[44] Nippon Suisan, however, has continued to focus on fish products. It maintained its supply through more concentrated fishing on the open sea, purchase of products from foreign fisheries, and the establishment of new joint ventures in key countries such as the United States, Canada and Russia.[45] Whereas the proportion of profit contributed by the fishing division of the biggest seven fishing companies dropped from 37 percent in 1975 to 22 percent in 1986, the import divisions have been upgraded and are expanding.[46] They are able to guarantee a stable supply to the Japanese market through financial control over factory vessels, such as those US vessels operating in Alaskan waters, as well as heavy investments in processing plants abroad.[47]

Conclusion

Although Japan was quite successful in its search to secure guaranteed access to important fishing grounds in the early post-war period, the situation worsened for distant-water fisheries with the application of greater fisheries restrictions worldwide. Despite a changing environment that included increasing imports and falling distant-water catch rates, the Japanese government continued to adhere to the policy goal of food security and self-sufficiency. The continuance of this policy is largely the result of the specific interests that the influential rice industry has in maintaining the food control system, which is based upon self-sufficiency.

Notes

1 Reiko Niimi, 'The Problem of Food Security', *Japan's Economic Security*, Nobutoshi Akao, ed. New York: St Martin's Press, 1983, 169.
2 Ibid., 169–70.
3 John W.M. Chapman, R. Drifte and I.T.M. Gow, *Japan's Quest for Comprehensive Security*. London: Frances Pinter, 1983, xi–xviii.
4 Yuichiro Nagatomi, *Masayoshi Ohira's Proposal: To Evolve the Global Society*. Tokyo: Foundation for Advanced Information and Research, 1988, 228–30.
5 Ibid., 228.
6 Food security was identified along with four other 'security tasks' that were deemed the core of comprehensive security policy: 1 promoting Japan-United States relations 'which in terms of military cooperation are more concrete and in terms of overall cooperation are more comprehensive'; 2 strengthening defence capability; 3 improving the management of relations with the People's Republic of China and the Soviet Union; and 4 achieving energy security. Ibid., 241.
7 Ibid., 258.
8 Ibid., 258.
9 Ibid., 261.
10 Ibid., 227.
11 Niimi, 'The Problem of Food Security', 186.
12 Naomi Saeki, 'Development of Trade in Agriculture Products and Border Adjustment in Agriculture', *Agriculture and Agricultural Policy in Japan*, IAAE Conference, ed. Tokyo: Tokyo University Press, 1991, 121.
13 Fred Sanderson, *Japan's Food Prospects and Policies*. Washington, DC: The Brookings Institution, 1978, table 2–5, 13.
14 Ibid., 13.
15 Sanderson, *Japan's Food Prospects and Policies*, 5–6.
16 Ibid., 11.
17 Toshiyuki Kako, 'Economic Development and Food Security Issues in Japan and South Korea', *Food Security in Asia: Economics and Policies*, Wen Chern, Colin Carter and Shun-Yi Shei, eds. Cheltenham, UK: Edward Elgar, 2000, 121.
18 Sanderson, *Japan's Food Prospects*, 7.
19 Ibid., 7.
20 Yujiro Hayami, *Japanese Agriculture under Siege: The Political Economy of Agricultural Policies*. Hampshire: Macmillan Press, 1988, 11.
21 Aurelia George Mulgan, *The Politics of Agriculture in Japan*. London: Routledge, 2000, 240. Nōkyō has acted as government-designated rice collection agents for more than two-thirds of all collectors, while an affiliated agency, Zennō, served as the primary seller of rice to the government with a nearly 100 percent share. The cooperatives also received from the government, through the Domestic Rice Control Account of the Food Control Special Account (FCSA) of the national General Account Budget, various subsidies such as delivery fees, fees for rice storage, interest subsidies for advance payments of rice, incentive and marketing assistance payments, fees for rice inspection as well as marketing commissions from farmers. Agricultural cooperatives were thus able to capture most of the economic benefits associated with a closed, subsidized and price-supported rice market. Ibid., 241–45.
22 Ibid., passim.

23 Path dependency is also shaped by the conditions of increasing returns. Although a number of possibilities may exist at the outset of the adoption of a certain policy, positive feedback can lead to a 'single equilibrium' that is resistant to change. When there is a high start-up cost in establishing new political/social institutions, the cost of change is just as high with the additional costs of changing entrenched behaviours, altering expectations and modifying the coordination required among various institutions or technologies. This leads to a situation whereby the status quo is quite resilient and the costs involved in change are often prohibitively high. Paul Pierson, 'Increasing Returns, Path Dependence, and the Study of Politics', *American Political Science Review* vol. 94 (June 2000): 251–67.

24 Communication with Director, International Fisheries Division, Economics Bureau, MOFA, 1 August 2005.

25 Tsuneo Akaha, *Japan in Global Ocean Politics.* Honolulu: University of Hawaii Press, 1985, 140.

26 Organisation for Economic Co-operation and Development (OECD), *Review of Fisheries in OECD Countries: Country Statistics 1999–2001*. Paris: OECD, 2003, 254.

27 Liberal Democratic Party, *Manifesuto* (Manifesto), National Election Campaign, November 2003.

28 Michael Weber, 'Effects of Japanese Government Subsidies of Distant Water Tuna Fleets', *Subsidies and Depletion of World Fisheries: Case Studies*, World Wildlife Fund, ed. Washington, DC: World Wildlife Fund, 1997, 121–22.

29 Ibid., 122.

30 Ibid., 122.

31 There is a discrepancy in the ratio of consumption (i.e. they add up to more than 100 percent) between domestic production and imports as a result of some of these products being used for export or for non-consumable use, such as fertilizer.

32 Officials at the Fisheries Agency were not able to offer an explanation for this jump in price. One possible explanation may be that salmon catches were being increasingly restricted on the high seas during this time and their import was substantially higher in cost than domestic production.

33 Glenn Hook, Julie Gibson, Christopher Hughes and Hugo Dobson, *Japan's International Relations: Politics, Economics and Security.* London: Routledge, 2001, 113.

34 Hasegawa Akira, 'Kokusaika-jidai no nihon-suisangyō no dōkō to kadai: yunyū mondai wo chūshin ni shite' (Trends and Issues for Japanese Fisheries in an Age of Internationalization: Focus on the Question of Imports), in *Sekai no naka no nihon gyogyō* (Japanese Fisheries in a Global Perspective), Yasuhiko Taki, ed. Tokyo: Seisandō, 1993, 10.

35 The most numerous of joint ventures have been the non-equity, contractual kind, which provided for over-the-side deliveries from coastal state catchers to Japanese factory vessels. See Olav Stokke, 'Transnational Fishing: Japan's Changing Strategy', *Marine Policy* vol. 15 (July 1991).

36 Kenji Horiguchi, 'Suisan-bōeki no kōzō to mondai-ten' (The Structure of Fisheries Trade and Related Problems), *Gyogyō Keizai Kenkyū* (Fisheries Economics Research) vol. 31 (January 1987): 70.

37 Akira, 'Kokusaika-jidai no nihon-suisangyō', 18.

38 Interview with Nishimura Masashi, Manager, International Section, Japan Fisheries Association, 15 December 2000.

39 Kazutoshi Kase, 'Globalization and Sustainability of Small-scale Fisheries', unpublished paper presented at ICFO, International Symposium on the Sustainability of Small-Scale Fisheries and Fishing Communities, held in Bangkok, 27 January 2005.

40 Interview with Director, International Fisheries Division, Economics Bureau, MOFA, 26 February 2004.
41 Stokke, 'Transnational Fishing', 234.
42 Ibid., 234.
43 'Taiyo Fishery Co., Ltd', *International Directory of Company Histories, Vol. 2.* New York: St James Press, 1997, 579.
44 Ibid., 579.
45 Ibid., 'Nippon Suisan Kaisha Limited', 552.
46 Stokke, 'Transnational Fishing', 241.
47 Ibid., 241.

Bibliography

Akaha, Tsuneo. *Japan in Global Ocean Politics.* Honolulu: University of Hawaii Press, 1985.

Chapman, John, R. Drifte and I.T.M. Gow. *Japan's Quest for Comprehensive Security.* London: Frances Pinter, 1983.

George Mulgan, Aurelia. *The Politics of Agriculture in Japan.* London: Routledge, 2000.

Hayami, Yujiro. *Japanese Agriculture under Siege: The Political Economy of Agricultural Policies.* Hampshire: Macmillan Press, 1988.

Hook, Glenn, Julie Gibson, Christopher Hughes and Hugo Dobson. *Japan's International Relations: Politics, Economics and Security.* London: Routledge, 2001.

Kako, Toshiyuki. 'Economic Development and Food Security Issues in Japan and South Korea', in *Food Security in Asia: Economics and Policies*, Wen Chern, Colin Carter and Shun-Yi Shei, eds. Cheltenham, UK: Edward Elgar, 2000.

Nagatomi, Yuichiro. *Masayoshi Ohira's Proposal: To Evolve the Global Society.* Tokyo: Foundation for Advanced Information and Research, 1988.

Niimi, Reiko. 'The Problem of Food Security', in *Japan's Economic Security*, Nobutoshi Akao, ed. New York: St Martin's Press, 1983.

OECD (Organisation for Economic Co-operation and Development). *Review of Fisheries in OECD Countries: Country Statistics 1999–2001.* Paris: OECD, 2003.

Saeki, Naomi. 'Development of Trade in Agriculture Products and Border Adjustment in Agriculture', in *Agriculture and Agricultural Policy in Japan*, IAAE Conference, ed. Tokyo: Tokyo University Press, 1991.

Sanderson, Fred. *Japan's Food Prospects and Policies.* Washington, DC: The Brookings Institution, 1978.

Stokke, Olav. 'Transnational Fishing: Japan's Changing Strategy', *Marine Policy* vol. 15 (July 1991): 231–43.

7 Comprehensive security in action
International diplomacy and negotiations

As mentioned in earlier chapters, new normative developments that suggested an increasingly restrictive international environment with respect to fisheries management required Japan to apply an alternative approach for finding and securing fish resources. This new strategy was divided into four main components involving: developing coastal fisheries more intensively; increasing the amount of international imports into the domestic market; negotiating new bilateral agreements to permit access to host countries' fishing grounds; and promoting the notion of increased or free access to fisheries resources at multilateral forums and organizations. Whereas the first two components of this strategy were discussed in greater detail previously, this chapter will examine the latter two policy initiatives which tested Japan's diplomacy on the international stage.

Apart from the domestic efforts discussed in the last chapter, Japan was also required to renew its foreign policy agenda in order to renegotiate bilateral agreements in light of the changes to its own domestic law and the international oceans regime with the conclusion of a United Nations Convention on the Law of the Sea (UNCLOS) agreement in 1982. The first priority was the renegotiation of bilateral fisheries agreements with Japan's neighbours: the USSR (and later, Russia), South Korea and the People's Republic of China (PRC). Japan also negotiated new bilateral agreements elsewhere, as will be seen in the case of tuna access agreements negotiated in the South Pacific. In the case of developing countries, official development assistance (ODA) played a large role in shaping negotiation strategies and producing outcomes favourable to Japan's distant-water fishing industry.

Finally, Japan sought to promote increased access to fisheries resources at multilateral forums and organizations. As a member of various Food and Agriculture Organization (FAO) treaties and institutions, Japan has abided by new rules governing fisheries management principles, although it is less forthcoming with respect to the application of the precautionary principle. Japan has used its negotiating power to press for recognition of the rights of distant-water fishing nations, which have typically been in competition with the rights of coastal states regarding the application of international oceans law. Japan is also a member of several multilateral fisheries organizations and a study of Tokyo's negotiating stance within the Convention for the Conservation of Southern Bluefin Tuna (CCSBT) will be used to illustrate the kinds of policies and positions adopted for the opening and expansion of a sustainable fishery deemed vital by Japanese interests. In general, Japan seeks to counter measures that limit fishing effort and restrict fishing grounds while promoting the cause of expanded access to sustainable fisheries worldwide.

Negotiating new bilateral agreements

As was mentioned in a previous chapter, the People's Republic of China and South Korea were exempted from the fisheries jurisdiction Japan declared in 1977. Increasing numbers of fisheries disputes between foreign trawlers and Japanese coastal fishermen, however, signalled the need to renegotiate new fisheries agreements between Japan and its neighbours. Fishing fleets from South Korea and the PRC had grown considerably in size during the 1980s and were in direct competition with Japanese fishermen, even, at times, in fishing grounds close to Japan's western and southern coastlines.[1] Moreover, Japanese coastal fishermen were in constant fear of a renewed fishing effort by their greatest fishing competitor to the north, the Russians.

Although earlier bilateral fisheries agreements were designed to provide reciprocal access to fishing grounds, Japanese distant-water fleets were the main beneficiaries since they were allowed access to fishing grounds near South Korea, the PRC and Soviet/Russian territories. South Korea and the PRC did not exercise their right to fish in Japanese waters due to their limited fishing capacity. As Korean and Chinese fishing fleets started going further afield into Japanese waters in the 1990s, however, the Japanese government sought to negotiate new agreements to limit their access and prevent competition with Japanese coastal fishers.

Japan also needed to respond to the worldwide trend toward the declaration of exclusive economic zones (EEZs) in order to claim exclusive access to living and mineral resources in ocean zones contiguous to its territory. Earlier bilateral agreements had freedom of the high seas as their central tenet but now new agreements needed to be concluded that employed the new rights conveyed by UNCLOS. On the basis of earlier bilateral agreements and Japan's 1977 fisheries law, Japanese waters beyond the 12-mile territorial sea remained high seas for Chinese and Korean trawlers, prompting these fishermen to overexploit fish stocks in grounds contiguous to the Japanese coast. The Japanese government wished to lay claim to these living resources and protect its own coastal fishers from undue competition.[2]

As a result of these factors, the Japanese government ratified UNCLOS in 1996 and subsequently promulgated the 'Law on the EEZ and the Continental Shelf' in June 1996, which entered into force in July 1996.[3] The law provides for the regulation of catches and the supervision of fisheries within Japan's EEZ. It also sought to establish new fisheries boundaries around Japan, extending up to 200 nm on its south-eastern coasts. Since a similar declaration on its western and northern boundaries would overlap with the jurisdictions of neighbouring states, Japan was required to negotiate with these countries under Article 74(3) of UNCLOS to establish a provisional boundary.[4]

Soviet/Russia bilateral agreements with Japan

The 1956 fisheries treaty between Japan and the Soviet Union/Russia was modified with three new fisheries agreements after the introduction of EEZs in the mid-1970s. The first of these agreements, the Japan-Russia Reciprocal Fisheries Agreement, was signed and brought into force in December 1984. It defined a provisional boundary for EEZs between the countries on Japan's northern waters and also established a joint fisheries council that made decisions regarding the application of treaty provisions, environmental status of fish stocks, allocations of fish quotas (especially with regard to snow crab and salmon catches) and other practical matters regarding the management of ocean resources.[5] On the basis of this agreement, fish catches have been allocated to Japan in one of two ways: fish quotas have been divided equally by both the Soviet Union/Russia and Japan; or Japan has paid access fees to obtain part of the Soviet/Russian quota.[6] In general, Japan's allocation within the Soviet/Russian EEZ has fallen since the imposition of EEZs due, in part, to the declining state of Alaskan pollack stocks and to the replacement of Japanese fishing effort with Russian ventures.[7]

Two other important fishing agreements were concluded with the Soviet Union/Russia in 1985 and 1998. The first of these was the Japan-Soviet Fisheries Cooperation Agreement of May 1985, which sought to provide joint management of salmon originating from Soviet/Russian rivers. It established quotas for salmon fisheries and sponsored scientific studies for the protection of salmon populations.[8] This agreement also provided for annual bilateral meetings to establish salmon quotas on a yearly basis. The second fisheries agreement was concluded in February 1998. Called the Operation Framework Agreement for the Waters Surrounding the Northern Islands, this treaty provided for the joint management of fishing resources near the disputed islands north of Japan. This agreement established annual meetings at the governmental and unofficial levels to decide upon quotas and fees for these fisheries without prejudicing Japan's claims to the islands.[9]

1997 Japan-PRC fishery agreement

The 1955 Japan-People's Republic of China joint fisheries agreement, which was discussed in Chapter 3, continued to govern fisheries relations between the nations until it was modified in 1975. After the normalization of relations between Japan and the PRC in 1972, the two nations opted to replace their previous private arrangement with a formal government-to-government agreement. They also added a joint fisheries council that discussed management issues within jointly managed zones. However, the 1955 agreement remained the same in all other aspects.[10]

In November 1997, Japan and the PRC concluded a new fishery agreement of substantially different proportions based on the provisions of UNCLOS. This agreement entered into force in February 2000. Both parties agreed provisionally to extend EEZ waters to 52 miles from their baselines and, in areas where it was difficult to distinguish each party's EEZ boundary, provisional waters that were jointly managed were established in an effort to avoid conflicting claims.[11] A Joint Fishery Committee (JFC), similar in function to the 1975 agreement, was established to manage fisheries within the provisional waters, and China and Japan agreed to give each other's nationals the right to fish in each other's EEZs in accordance with the principle of reciprocity, subject to domestic legislation and decisions made by the JFC.[12]

1999 Japan-South Korea fishing agreement

Talks between Japan and South Korea were initiated in 1996, but took over three years to arrive at an agreement. One of the main points of contention between the two nations was the status of an island called Takeshima (Dokdo in Korean) and the fishing rights that would be enforced around it. A compromise solution was found by declaring the waters around this island 'middle waters', which would be jointly managed by both Japan and South Korea.[13] Within this zone each nation would be responsible for managing the flag ships of its respective country and would forfeit the right to arrest the fishermen or board ships of the other.

The new fishing agreement was concluded in November 1998 and entered into force in January 1999. It stipulated that EEZ waters would be extended to 35 miles from the baseline of each country and a joint fisheries committee, the South Korea-Japan JFC, would be established to make decisions about the conditions of fishing grounds, fishing licensing, gear types, enforcement and other administrative matters.[14] The JFC is currently composed of one commissioner and one representative from each country. Moreover, catch quotas are made approximately the same for each country within shared waters.[15]

The management of fisheries within the 'middle waters', however, has proven contentious enough that the status of the fisheries agreement may be in jeopardy. Japan has argued recently that Korean vessels are overfishing within this zone, particularly for highly valued snow crab, and that illegal fishing is running rampant to the extent that Japanese fishermen have been crowded out of these fishing grounds.[16] Japanese negotiators have been pushing for the establishment of a joint management council that would be empowered to set fishing quotas, seasonal limits and gear regulations on these fisheries, but the Korean government has so far rejected all such proposals.[17] Flare-ups in March 2005 over whether Japan or South Korea controls the

Takeshima Islands also complicated these discussions, since many of the most productive fishing grounds in the 'middle waters' are found near there.

Japanese negotiators have been able to achieve one common goal in recent years: the gradual reduction of fishing effort within Japan Sea fishing grounds. Whereas catch allocations in 1999 favoured South Korea with an allotment of 148,000 tons of fish versus Japan's 93,000 tons, by 2005 this figure had been reduced to 67,000 tons for each country.[18] Parity in fish catch allotments has been realized in addition to a reduction in quotas as preferred by the Japanese Fisheries Agency.

The bilateral fisheries treaties negotiated with Russia, the PRC and South Korea shared several distinctive characteristics. Since Japanese negotiators could no longer base their fishing claims upon the principle of freedom of the seas, they had to offer new incentives to neighbouring nations in order to conclude desired fishing treaties. Thus, each agreement reaffirmed Tokyo's commitment to reciprocal access to fisheries in spite of the potential for conflict this created with domestic Japanese fishers. Moreover, Japanese negotiators bypassed potential impediments inherent in territorial disputes by agreeing to manage fisheries jointly in waters that were disputed. Finally, Japan promised to hold bilateral meetings with the other signatories on an annual basis in order to set quotas and to discuss fisheries regulations and restrictions.

Fisheries aid in the bilateral negotiations of South Pacific fisheries

The Japanese government has also played an active part in assisting members of its fishing industry with access to valuable tuna fishing grounds within coastal states' 200-nm zones in the South Pacific. Governmental assistance to the distant-water fishing industry is clearly seen in the difficult tasks of negotiating bilateral agreements with coastal states on behalf of fishing companies, especially when such talks are leveraged with promises of ODA in the case of poorer states.

Fisheries aid, as one component of ODA, is managed by several agencies in Japan, the two most important of which are the Overseas Fisheries Cooperation Foundation (OFCF) and the Japan International Cooperation Agency (JICA). OFCF is a non-profit Japanese government organization that was established in June 1973 and is administered by the Fisheries Agency.[19] It assists the Japanese private sector while strengthening international relations between Japan and recipient countries. The OFCF 'aims to promote the development of fisheries in the coastal countries with which Japan has close relationships in the fisheries field and to effectively manage international marine resources within regional fisheries organizations in order to enhance amicable fisheries relationships through technical and economic cooperation'.[20] The OFCF has a budget of around US$28–$34 million.[21]

Whereas the OFCF largely provides funds for equipment and services, JICA's primary method of aid is technical assistance. Although JICA is mostly administered by the Ministry of Foreign Affairs (MOFA, by which the president is appointed), the Ministry of International Trade and Industry (MITI) and the Ministry of Agriculture, Forestry and Fisheries (MAFF) also appoint some executive managers on rotation and have an influence on decision making as well.[22] JICA's primary role is to carry out basic design services and to help implement grant aid projects, including fisheries projects related to technical cooperation. In 2004, JICA helped MOFA to administer a 60 billion yen (US$520 million) fisheries aid budget.[23]

Fisheries aid is offered to developing states to assist economic capacity for catching, processing and retailing fish and shellfish. Such aid takes the form of fisheries services and goods, including ships for resource assessment, monitoring and training; cold-storage and ice-making facilities; fish hatcheries and processing centres; and port facilities.[24] Training, scientific studies and study trips to Japan are also offered as ways to develop human resources for fishing. According to Figure 7.1, fisheries grant aid has almost consistently represented nearly 5 percent of the overall grant aid budget of Japan.

Japan has used ODA aid to facilitate access to important fishing stocks in the period following the expansion of EEZs by coastal states. This is most clearly evident in the South Pacific, where ODA has been used to secure access to valuable tuna stocks that migrate though the 200-nm EEZs of many

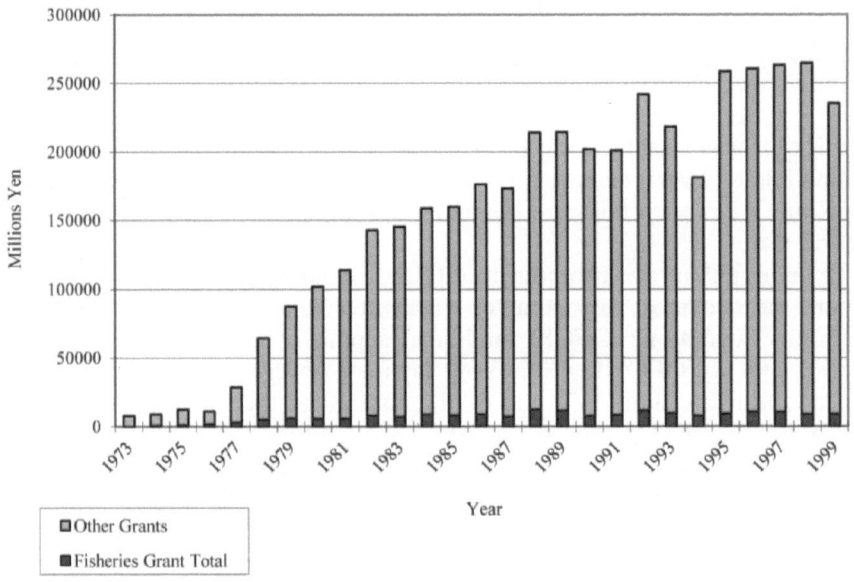

Figure 7.1 Japanese aid grants, 1973–99
Source: Ministry of International Trade and Industry, *Keizai kyōryoku no genjō to mondaiten* (The Conditions and Problems of Economic Cooperation). Tokyo: Keizai sangyō-shō, 1973–2000

Pacific islands. Whereas few Japanese officials will admit to such a practice, there is much evidence in support of the use of aid to help subsidize access fees and to guarantee Japanese fishing boats access to tuna fishing grounds.

At a time when tuna and skipjack were valued as exports and becoming increasingly important as imports for a valuable domestic market, the Pacific islands were key suppliers of tuna to the Japanese market. In 1977, EEZs of fifty-four countries covered fishing areas that produced 48 percent of Japan's tuna production (and 41 percent of skipjack), for which areas the Pacific islands held a significant share.[25] Pacific Ocean fishing grounds, which include those grounds within 200 nm of Pacific island states, provided more than 90 percent of Japan's tuna supply in 1980.[26] Consequently, pelagic fisheries have been a key source of foreign exchange for many Pacific island states, including Kiribati, Tuvalu, Marshall Islands, Federated States of Micronesia, Fiji, Tonga and Solomon Islands.[27] This situation was amplified after many Pacific islands declared their own EEZs that enclosed tuna fishing grounds valuable to the Japanese, and as fishing grounds in the Pacific became increasingly attractive due to their proximity to Japan and the increasing appetite for fresh fish for the sashimi market.

Most Pacific island states declared their own 200-nm fishing zones between 1977 and 1979. These islands were able to claim administrative jurisdiction over ocean areas that were thousands of times the size of their land area. For example, the EEZ declared by the Cook Islands allowed them to claim a fishing area of 1.8 million square km whereas their actual land area was a mere 237 square km.[28] This created a new geopolitical situation in the Pacific, requiring Japan to negotiate access agreements with island states in order to catch or procure the tuna fish needed for the domestic market.

The Japanese government played a supportive role in negotiating access agreements for private companies largely through the use of fisheries grant aid.[29] In her study of ODA and tuna access agreements in the South Pacific, Sandra Tarte found a positive correlation between the levels of fisheries grant aid and the number of access agreements to be found in Pacific islands. She concludes that 'fisheries grant aid was introduced to support Japan's distant-water fishing industry in foreign fishing grounds enclosed by coastal state exclusive economic zones'.[30] The OFCF sought to promote fisheries cooperation by channelling government funds to the private sector to create commercial networks with coastal states, particularly in the form of joint ventures, to the extent that OCFC funded up to 70 percent of capital investments in joint ventures.[31]

Fisheries grant aid, in the form of goods and services offered to Pacific island nations, is considered by the Fisheries Agency and the Japanese government as part of the fee charged for access to tuna fishing grounds.[32] In this sense, fisheries grant aid can also be considered an informal subsidy to the distant-water tuna fisheries, serving as a conduit for government payments to the private industry. As stated by David Doulman, 'Japan views the goods and services supplement as being part of the cost of access, although FFA

[Forum Fisheries Agency] member countries take a different view, because the supplement is not paid in cash. Because 75 percent of the goods and services contribution is paid by the Japanese government, the supplement represents a subsidy to the fishing industry'.[33] The amount of money the Japanese government is paying for access agreements is substantial relative to the amount paid by private fishing companies. The Japanese government, through the OFCF, paid US$20 million per year between 1982 and 1992 compared to US $14.5 million paid by industry for Japanese access fees.[34]

Access fees and fisheries aid grants represent a significant portion of the overall budgets of some Pacific island countries. In Kiribati, for example, access fees accounted for 9 percent of the government revenue in 1981, rising to 25 percent in 1986 as a result of a new agreement with the Soviet Union.[35] Aid also plays an important role in the region's governmental budgets, with the ratio of aid to gross domestic product (GDP) being more than 20 percent on average and constituting as high as 80 percent in some cases.[36] Some 70 percent of the total grant aid to the Pacific islands between 1975 and 1986 was fisheries grant aid.[37]

In promoting the link between fisheries access and ODA, the Japanese industry and government officials have been successful in securing a steady supply of tuna for the Japanese market, despite the introduction of EEZs in the late 1970s. Whereas some South Pacific representatives have argued that such aid linkages lead to a relationship of dependence, Japanese industry and government officials applaud the system for creating fishing capacity and facilities in poorer South Pacific countries as well as ready market access for sashimi quality tuna for Japanese consumption. Nonetheless, there is growing debate whether South Pacific countries would see better returns if they were to charge access fees at market rates, or harvest tuna themselves, rather than accept ODA as an unofficial subsidy to access fees.[38]

Promoting free or increased access to fisheries resources at multilateral forums

The final strategy Japan used to counter the effects of the appropriation of its overseas fishing grounds was to promote free access to responsible fisheries on the high seas as well as to minimize the application of restrictions on important distant-water fisheries at various international forums and commissions. Examples of these strategies will be examined with respect to Japan's adherence to the FAO's Code on Responsible Fisheries and the subsequent campaign against illegal, unreported and unregulated (IUU) fisheries, as well as Japan's negotiation stance and policies with respect to the CCSBT.

During the early 1990s, several new international instruments were under negotiation to help further define, extend and apply the 1982 Law of the Sea. These negotiations resulted in several new agreements being applied for anadromous species, the rights and obligations of coastal and flag states,[39] as well as a new instrument governing the high seas, as discussed in a previous chapter. The rights and obligations of coastal versus distant-water fishing states were being formulated at this time, picking up where UNCLOS III negotiations had left off.

The Fish Stocks Agreement of 1995 set fairly restrictive rules on fishing for highly migratory and straddling fish stocks. In particular, the application of the 'precautionary approach' prohibited fisheries when there was reasonable, but not necessarily scientific, proof that the future status of the fish stocks was in danger. This placed the onus on distant-water fishing nations, such as Japan, to provide a heavy burden of proof to show scientifically that their fishing activities were not harmful.

Code of conduct

A new Code of Conduct for Responsible Fisheries was negotiated under the direction of the FAO during the same year as the Fish Stocks Agreement. As a result, these two international instruments complement each other in several ways. One important difference, however, is that whereas the Fish Stocks Agreement is binding upon signatory states, the Code of Conduct is strictly voluntary. It is meant to serve as a guide for how fisheries should be conducted to ensure the future sustainability of fisheries and to minimize the negative impact upon the marine environment. The code serves as the basis of the strategic framework of the FAO's Fisheries Department.[40]

The code sets out international standards of conduct and principles for responsible fisheries with a view to ensuring the effective conservation, management and development of living aquatic resources, with due respect for the ecosystem and biodiversity. The code states as a fundamental principle that the right to fish carries with it the obligation to do so in a responsible manner so as to ensure effective conservation and management, and that states should prevent overfishing and excess fishing capacity.[41] As with the Fish Stocks Agreement, the code also assigns particular emphasis to the precautionary approach.[42] It also encourages fishing states to comply with international conservation and management measures, apply effective enforcement, and establish appropriate mechanisms for the monitoring and control of fishing vessels.[43]

In order to encourage the application of these principles, the FAO has elaborated four sets of 'action plans' on certain priority areas. These include the International Plans of Action for Reducing the Incidental Catch of Seabirds in Longline Fisheries, for the Conservation and Management of Sharks, for the Management of Fishing Capacity, and to Prevent, Deter and Eliminate Illegal, Unreported and Unregulated (IUU) Fishing.[44] These plans of action are also voluntary, but call upon states to take detailed measures, often within a specified timeframe, and such national policies are periodically reviewed by the FAO's Committee on Fisheries (COFI).[45]

Japan has widely endorsed the action plans and has developed several policies in compliance. It has taken an especially strong position with respect to controlling IUU fishing and was responsible for pressuring the FAO to adopt the action plan on this topic.[46] Japan's commitment in this regard is important since it happens to be the world's largest market for fish products as well as fish captured by IUU fishing techniques. It attempts to deter and eliminate such practices mainly through a ban on the importation of IUU fish products into Japan, as well as requiring vessel documentation and catch data on fish catches brought into Japanese ports.[47]

Adopting such a hard-line stance with respect to IUU fishing appears to serve two functions. First of all, it ensures that management efforts and conservation practices by law-abiding fishermen are not undermined by those who fish illegally, or do not report their catches or submit to fisheries regulations. Second, it protects domestic fishermen who are subject to management regulations from having their profits undercut by foreign fishermen who use less costly equipment and methods outside regulatory control.[48]

Whereas Japan has strongly supported the code's actions plans, it has been much less enthusiastic about the application of the precautionary principle that favours the fishing interests of coastal states over distant-water fishing nations. Perhaps as a reaction to the rising support of coastal state rights with respect to ocean law, Japan elected to co-host a conference with the FAO on the subject of the important role played by fisheries in food security.[49]

Hosted in Kyoto in December 1995, the International Conference on the Sustainable Contribution of Fisheries to Food Security issued a multilateral statement, called the Kyoto Declaration, which stated that fisheries have an important role to play in the provision of natural renewable resources for human consumption and unless urgent action is taken, the future supply of fish and fishery products will not be able to meet growing demand.[50] It also stated that fisheries play a significant role in providing food security, both through food supply and through economic and social well-being, and that there should be respect for social, cultural and economic differences among states and regions in the use of living resources, especially cultural diversity.[51]

The Kyoto Declaration helped to codify Japan's position at various international conferences over the years, in effect saying that fisheries play a unique and integral role in the food security of the Japanese nation, as with developing nations, and that this position should be respected in the future notwithstanding the rights and obligations imposed by other multilateral agreements. The Kyoto Declaration serves as an instrument in international law, although non-binding, which supports Japan's claims for broader access to the world's fisheries, including the resumption of whaling that serves an example of Japan's 'cultural diversity'.[52]

As a member of various multilateral fisheries conventions and treaties, Japan has also promoted open access to responsible fisheries and attempted to minimize the application of restrictions on important distant-water fisheries. Japan is a participating member on many international fisheries organizations that govern fishing areas or specific stocks of fish, including the Indian Ocean Tuna Commission (IOTC), the International Commission for the

Conservation of Atlantic Tuna (ICCAT), the North Atlantic Fisheries Organization (NAFO), the International Whaling Commission (IWC) and the CCSBT, among many others.

Southern bluefin tuna

An examination of Japan's policy with respect to the Convention for the Conservation of Southern Bluefin Tuna will provide a useful illustration of Tokyo's larger fisheries policy in many other international fisheries forums. In particular, Japan has faced an increasingly restrictive environment with respect to cost of operations and increasing scarcity of fisheries resources, while also attending to the needs of a domestic industry that is trying to meet the burgeoning demand of the domestic market and face competition from such countries as South Korea and Taiwan.

Southern bluefin tuna (SBT) are a high-value species for the Japanese sashimi market. It is a highly migratory species that ranges from the South Atlantic to the Tasman Sea and the Indian Ocean, spawning in a single ground south of Java, Indonesia.[53] Since the SBT is a slow-growing species, not achieving maturity until about eight years and living as long as twenty years, it is highly vulnerable to overfishing. In recent years, intensive fishing of juveniles has led to drastically reduced adult populations and lower recruitment rates (the rate at which juveniles join the reproductive stock), which in turn has resulted in historically low stock levels with the potential of parent stock collapse as recruitment declines further.[54]

Japanese fishing fleets have been fishing SBT since 1952, following the lifting of the MacArthur restrictions. The Japanese fleet increased its catch from 556 tonnes in 1952 to a post-war peak of more than 77,000 tonnes in 1961.[55] Despite the expansion of the Japanese longline fleet in the 1960s due in part to high product prices and the low cost of fuel, Japanese catches hereafter declined from an average of 46,000 tonnes per year in the 1960s to 14,000 tonnes in 1987.[56] Increasing fishing effort that results in declining catch rates is a clear symptom of overfishing and there were many ensuing calls for greater catch restrictions in an effort to rebuild fish stocks. Japan now maintains a smaller, but active fleet targeting SBT in the Southern Ocean and in waters around Australia and New Zealand.[57]

In order to conserve and reduce fishing pressure on SBT stocks, the CCSBT was established in 1993 as an agreement among Australia, New Zealand and Japan. Although informal management arrangements existed prior to this, the CCSBT formalized them by providing for yearly meetings of a newly established commission and for joint scientific studies.[58] The CCSBT determines management and conservation measures for the fishery whereas the scientific committee advises the commission on stock population data and acceptable catch rates.[59] It should also be noted that the convention has a provision for other countries to join the agreement, enabling them to receive an SBT quota in exchange for adopting conservation measures. Due to lobbying and pressure applied by Japan and the other signatories, South Korea joined the convention in 2001 and Taiwan in 2002.[60] Since these countries were responsible for a significant level of unreported and unregulated fishing in previous years, their accession to the agreement strengthens the overall effectiveness of conservation measures, while making competition fairer with domestic Japanese fishermen who abide by the agreement.

Perhaps the most effective provision contained within the convention relates to mandatory quota-setting procedures and the allocation of total allowable catch among convention parties. In deciding quota allocations, the commission considers a number of factors. It takes advice offered by the scientific committee and considers the need for orderly and sustainable development of the fisheries, the interests of the coastal states through whose EEZs the SBT migrate, the interests of distant-water fishing nations that have historically fished SBT, and the contribution of each party to conservation and the enhancement of scientific research.[61] The most important provision for Japan, by far, is the recognition of its traditional dependence and historic fishing rights of SBT, which aligns closely with the Japanese government's position that the only fisheries arrangements that can be considered legal and legitimate are ones that recognize the interests of distant-water fishing nations.

It should be noted that Japan had initially resisted attempts to enshrine quotas in the convention. It finally acceded to them, however, in return for recognition of its rights as a distant-water fishing nation in the tuna fisheries of the region. An earlier informal arrangement in the early 1980s between Australia and Japan provided for a suggested total allowable catch (TAC), but this was non-binding as Japan did not wish to be compelled to abide by restrictions.[62] Mounting scientific evidence in the 1980s, however, suggested that SBT stocks were being depleted and constituted a mere fraction of the size of 1960 populations. Australia increased pressure on Japan to agree to quotas, but Japan resisted in almost every instance. Finally, the Australian prime minister threatened in 1989 that a moratorium might be needed to conserve SBT stocks and Japanese negotiators subsequently agreed to a 50 percent quota reduction for the 1989 fishing year which placed limits on Japan's longline fleet.[63]

Similar reasoning may have led Japanese negotiators to accept quotas in the convention itself, since there was a concurrent movement in the Convention on International Trade in Endangered Species of Wild Fauna and Flora (CITES) to have North Atlantic bluefin tuna stocks listed as an endangered species which would have qualified them for special protection and possible moratorium.[64] Although this never materialized, Japan still felt it was safer to agree to a conservation convention with quota restrictions than to risk another possible listing of its most commercially valuable fish product for special trade protection.

One of the chief objectives of the CCSBT is to adopt conservation measures to encourage a return of the SBT population to 1980 levels by the year 2020. To do so, quotas are supposed to be set at levels low enough to encourage the replacement of the adult stock. Such assessments for recovery are based upon the estimated recruitment levels of the population, creating a great deal of variance – from rapid recovery to further substantial decline – depending upon which specific hypothesis is used to model uncertainties in the assessments.[65]

Since there is a great deal of variance among assessments used by the scientists of each national delegation, there is significant variance among the quotas proposed by each country. It is sufficient to note that Japanese scientists have rather high estimates of juvenile recruitment, and therefore stock recovery, whereas scientists from New Zealand and Australia are much more pessimistic, predicting a low probability of recovery.[66] Therefore, Japan has argued for an increase in SBT quotas whereas Australia and New Zealand aim for a reduction.

As a result of these conflicting scientific estimates and political views, CCSBT negotiations for the 1996/97 year broke down in disagreement. Whereas an allocation of 5,265 tonnes for Australia, 6,065 for Japan and 420 for New Zealand was agreed in previous years, Japan's delegation had argued at first for an increase in 1996 of an additional 6,000 tonnes, which it later reduced to 3,000 tonnes, based on its own population estimates.[67]

Australia and New Zealand strongly objected to this proposal, after which Japan modified its stance and argued for an increase in quota – the same amount as requested prior – for three years to support an 'experimental fishing programme' (EFP) that would aim to resolve the scientific uncertainty that existed among the countries. Both Australia and New Zealand rejected the experimental fishing proposal as reckless, unnecessary and unscientific.[68] Japan thereupon declared in 1998 that it would conduct the programme 'unilaterally', prompting Australia and New Zealand to respond that this would constitute an abrogation of the CCSBT.[69] Australia and New Zealand then appealed to the International Tribunal for the Law of the Sea (ITLOS) to put an end to Japan's unilateral EFP.

The Japanese government staked out several positions before the ensuing ITLOS arbitration tribunal held in May 2000. Most importantly, the Japanese government argued that the issue was, fundamentally, a scientific dispute rather than a legal one, and therefore UNCLOS did not have jurisdiction.[70] Rather, they argued, the issue should be solved within the CCSBT framework since it provided for a dispute resolution system. On the other hand, the Australian and New Zealand governments maintained that Japan's EFP was misdirected and that its design and analysis were fundamentally flawed. In their view, Japan's EFP did not justify what they saw as the significant increased risk to the SBT stock.[71]

They argued that the unilateral injunction of the EFP abrogated the CCSBT agreement and therefore asked that ITLOS place a halt on Japan's EFP. They argued that the EFP violated the precautionary principle, which states that caution must be used by countries in taking decisions about actions that entail threats of serious or irreversible damage to the environment while there is scientific uncertainty about the effect of such actions. The principle requires caution and vigilance in decision making in the face of such uncertainty.

The ITLOS arbitration tribunal finally ruled that it did not have jurisdiction to make a ruling on the dispute. The tribunal, however, did make a number of recommendations. It agreed with Australia and New Zealand by stating that CCSBT member states should refrain from conducting an experimental fishing programme, except with the agreement of the other parties or unless the experimental catch is counted against its annual national allocation.[72] It also recommended that SBT annual catches should not exceed the annual national allocations at the levels last agreed by the CCSBT parties, which was a recommendation in support of the application of the precautionary principle.[73] In ruling that it did not have jurisdiction, however, the tribunal agreed with Japan's position that the dispute should be resolved within the CCSBT itself and thus revoked the provisional ruling prohibiting the conduct of the EFP.

Despite the dispute over the EFP, Australia, New Zealand and Japan have managed to work together to introduce a number of conservation and management measures under CCSBT. Most importantly, the CCSBT has introduced a Trade Information Scheme to track the point of origin of SBT. The aim of the programme is to cut down on the number of IUU vessels that are catching SBT, which undermine efforts by member countries to manage the resource.[74]

As the largest market for SBT, Japan is a necessary participant in the scheme and its cooperation is essential for the reduction of flag of convenience vessels and for maintaining accurate statistics. Japan is concerned that its commitment to reducing the capacity of its fleet and the opportunity costs incurred by participating in conservation schemes is being negated by catches by non-parties (for example, Indonesia) and IUU methods. CCSBT also maintains a list of approved vessels that are permitted to fish SBT.[75] Finally, quotas have been successfully negotiated in years subsequent to the SBT dispute and in the 2004/05 season they amounted to 6,065 tonnes for Japan, 5,265 tonnes for Australia, 1,140 tonnes for South Korea, 1,140 tonnes for Taiwan and 420 tonnes for New Zealand.[76] These figures remain unchanged for the original signatories to the convention.

Access to high-valued species, such as SBT, remains a high priority for the Japanese government. This has been illustrated at international forums through a concerted effort to expand, or in the very least maintain, catch rates for Japanese fishing fleets while also promoting the goal of fisheries conservation. Japan lends its support to these organizations, even at times when their decisions may be opposed to Japanese interests, largely because such international bodies recognize and reinforce the rights Japan enjoys as a distant-water fishing nation, which also helps temper the global movement granting wider management control rights to coastal nations. The development of international management arrangements, such as the Trade

Information Scheme and the acceptance of scientifically determined fisheries quotas, also illustrates Japanese commitment to ensuring the viability of its distant-water tuna operations. By bringing other countries into the CCSBT framework, Japan hopes to ensure the future effectiveness of this multilateral fisheries agreement while not undermining its own efforts at restraint.

Conclusion

In an effort to attain the goal of comprehensive security as outlined in Chapter 6, the Japanese government has had to adopt a new set of strategies that relinquished direct control of fisheries production and relied much more heavily upon imports and temporary access arrangements. Japan employed four strategies to help guarantee a steady supply of fish products: 1 developing coastal fisheries more intensively; 2 manage the amount of international imports into the Japanese market; 3 negotiating new bilateral agreements to permit access to host countries' fishing grounds; and 4 promoting the notion of increased access or free access to fisheries resources at multilateral forums and organizations.

The fishing zone claimed by Japan in 1977 gave its domestic fisheries a vast expanse fifty times as large as its previous territorial sea had and thus initial efforts to expand fish catch were easily attainable in the short term. Japan also witnessed a significant rise in fisheries imports from countries whose fishing capacity was significantly expanded due to the declaration of their own fishing zones. In some cases, it can be said that these foreign fishermen replaced the Japanese in the production of fish that were destined for the Japanese market.

Japan was also required to renegotiate bilateral agreements in light of the changes to its own domestic law and the international oceans regime with the conclusion of an UNCLOS agreement in 1982. The first priority was the renegotiation of reciprocal bilateral fisheries agreements with Japan's neighbours, the USSR (and later, Russia), South Korea and the People's Republic of China, which delineated fishing boundaries, protected Japanese domestic fishermen and established procedures for the management of fish stocks. Japan also negotiated new bilateral agreements in the South Pacific and elsewhere, whereby ODA played a large role in shaping negotiation strategies and producing outcomes favourable to Japan's distant-water fishing industry.

Finally, Japan sought to promote increased access to fisheries resources at multilateral forums and organizations. Increasing pressure to restrict access to various fisheries worldwide has led Japan to counter such measures that limited fishing effort and fishing grounds by promoting expanded access to living maritime resources worldwide. Japan has proven an active participant in multilateral forums, such as the FAO and the CCSBT, especially when these international organizations recognize Japan's inherent special rights as a distant-water fishing nation that has a traditional dependency upon marine resources.

Such participation, however, has rarely occurred without controversy and conflict. A growing worldwide movement recognizing the superior rights of coastal states in the management and protection of maritime resources constantly threatens Japan's special interests as a distant-water fishing nation needing a secure supply of fishery resources. During the past three decades, Japan has defended its interests against the appropriation of maritime resources by other nations claiming coastal state precedence.

Of additional concern to Japanese officials – this time at the sub-national level – is the spreading influence of an environmental movement that threatens to place greater restrictions, controls and even moratoriums on Japan's fishing effort. This movement, which is especially influential in Europe and North America and has widespread support at the governmental level, seeks to restrict access to and protect various fisheries worldwide, much as the enclosure movement did in the 1970s.

This conflict is especially acute in the case of whaling, which serves as an issue area that helps to define the parameters between the special rights enjoyed by Japan as a fishing nation and the duties inherent in the global community to protect a threatened resource. It also highlights the role of epistemic norm formation in the creation of new international values and agreements. These issues will be examined in the next chapter.

Notes

1 Fisheries Agency, *Waga kuni no suisan gaikō ni tsuite* (The Fisheries Diplomacy of Japan). Tokyo: Fisheries Agency, 2003, 3.
2 Ibid., 3.
3 Ibid., 3.
4 United Nations Convention on the Law of the Sea. For the full text of the agreement, see www.un.org/Depts/los/convention_agreements/texts/unclos/closindx. htm.
5 Fisheries Agency, *Waga kuni no suisan gaikō ni tsuite*, figure 1-2-11.
6 Ibid., 5.
7 Interview with Director, International Affairs Division, Fisheries Agency, 23 February 2004.
8 Fisheries Agency, *Wagakuni no suisan gaikō ni tsuite*, figure 1-2-11.
9 Ibid., 6.
10 Joon-Suk Kang, 'The United Nations Convention on the Law of the Sea and Fishery Relations between Korea, Japan and China', *Marine Policy* vol. 27 (March 2003): 121.
11 Zou Keyuan, 'Sino-Japanese Fishery Management in the East China Sea', *Marine Policy* vol. 27 (March 2003): 133–34. Disputes that flared up in 2005 and early 2006 between the PRC and Japan regarding which country controls oil and gas drilling rights occurred within the Provisional Zone to the south of Japan near the Senkaku (Daoyutai in Chinese) Islands.
12 Ibid., 133–35.
13 Sun Pyo Kim, 'The UN Convention on the Law of the Sea and New Fisheries Agreements in North East Asia', *Marine Policy* vol. 27 (March 2003): 99.
14 Kang, 'The United Nations Convention on the Law of the Sea', 117.
15 Ibid., 119.

16 Interview with Manager, International Affairs Division, Fisheries Agency, 8 March 2005.
17 Ibid.
18 Ibid.
19 Overseas Fisheries Cooperation Foundation, publicity material.
20 Ibid.
21 Anthony Bergin and Marcus Haward, *Japan's Tuna Fishing Industry: A Setting Sun or a New Dawn?* New York: Nova Science Publishers, 1996, 73.
22 Interview with Officer, Fisheries Division, JICA, 13 February 2004.
23 Ibid.
24 Ministry of International Trade and Industry (MITI), *Keizai kyōryoku no genjō to mondaiten, Heisei 12-do* (The Conditions and Problems of Economic Cooperation, 2000). Tokyo: Keizai sangyō-shō, 2000, 7–8.
25 Yoshiaki Matsuda, 'Postwar Development and Expansion of Japan's Tuna Fishery', *Tuna Issues and Perspectives in the Pacific Islands Region*, David Doulman, ed. Honolulu: East West Center, 1987, 87.
26 Sandra Tarte, *Japan's Aid Diplomacy and the Pacific Islands*, Pacific Policy Paper 26. Canberra: Australia National University, 1998, 2.
27 Ibid., 12.
28 David Doulman, *Fishing For Tuna: The Operations of Distant Water Fleets in the Pacific Islands Region*. Honolulu: East West Center, 1986, 43.
29 Access agreements are usually calculated in one of two ways: a lump sum payment in cash or payments based on the number of vessels and trips to fishing grounds.
30 Tarte, *Japan's Aid Diplomacy*, 93, 96.
31 Ibid., 86.
32 Yoshiaki Matsuda, 'Changes in Tuna Fishing Negotiations between Japan and the Pacific Island Nations', *Resources and Environment in Asia's Marine Sector*, James Barney Marsh, ed. New York: Taylor and Francis, 1992, 51. The practice continues despite the calls by Pacific island states to desist. Being such a large market for tuna, Japan is able to wield influential leverage over bilateral relations with these countries, leaving them with little bargaining power to untie aid from access fees.
33 David Doulman, 'Japanese Distant Water Fishing in the South Pacific', *Pacific Economic Bulletin* vol. 4 (December 1990): 25–26.
34 Bergin and Haward, *Japan's Tuna Fishing Industry*, 74.
35 Roniti Teiwaki, 'Access Agreements in the South Pacific: Kiribati and the Distant Water Fishing Nations, 1979–86', *Marine Policy* vol. 11 (October 1987): 283.
36 Tarte, *Japan's Aid Diplomacy*, 11.
37 Ibid., 2.
38 One important study in this regard is Elizabeth Petersen, 'The Catch in Trading Fishing Access for Foreign Aid', *Marine Policy* vol. 27 (May 2003): 219–28. In this study, Petersen suggests that if access fees were maximized, there is a potential for the access fees to match, possibly double, total Japanese aid to the region. She argues that aid dependency is decreasing the transparency of fishing treaties, decreasing the flexibility of government spending, exposing the Pacific island countries to large financial risks associated with possible aid withdrawal, and is responsible for stifling the region's fisheries and broader economic development.
39 Flag state refers to the state to which a fishing vessel is registered. Fishing vessels on the high seas are free from monitoring, inspection, boarding and seizure except in the case when that state to which a vessel is registered requires one of its vessels to do so.
40 Interview with Fishery Resources Officer, Fisheries Department, FAO, 28 January 2002.

41 FAO, *Code of Conduct for Responsible Fisheries.* Rome: FAO, 1995, Art. 6.
42 Ibid., Article 7.
43 Ibid., Article 6.
44 FAO, *Code of Conduct*, www.fao.org/fi/agreem/codecond/codecon.asp. Much of the IUU fishing is done by flags-of-convenience vessels that register their boats in countries with non-existent or lax fisheries regulation and control, and then proceed to fish indiscriminately and with impunity in international waters.
45 Moritaka Hayashi, 'Three Decades' Progress in High Seas Fisheries Governance: Towards a Common Heritage Regime?' Paper presented at the Center for Oceans Law and Policy's Twenty Sixth Annual Conference, 'Stockholm Declaration and Law of Marine Environment', Stockholm, 22–25 May 2002.
46 Fisheries Agency, *Waga kuni no suisan-gaikō ni tsuite*, 16.
47 Interview with Director, International Fisheries Division, Economics Bureau, MOFA, 26 February 2004.
48 Ibid.
49 MAFF, *Update* no. 151, 22 December 1995.
50 FAO, Kyoto Declaration, www.fao.org.
51 Ibid.
52 *Asahi Shimbun*, 'Shokuryō anpo-jō no jūyō-na yakuwari wo hyōka – kokusai gyogyō kaigi-de kyōto sengen saitaku' (An Important Role in Food Security: The Adoption of the Kyoto Declaration at an International Fisheries Conference), 10 December 1995.
53 Tom Polacheck, 'Experimental Catches and the Precautionary Approach: the Southern Bluefin Tuna Dispute', *Marine Policy* vol. 26 (July 2002): 284.
54 Ibid., 284.
55 Bergin and Haward, *Japan's Tuna Fishing Industry*, 134.
56 Ibid., 134. Longlining is a fishing technique that utilizes baited hooks deployed from small branch lines (known as snoods) which are attached to a main line that can be over 120km in length.
57 Ibid., 134.
58 CCSBT Convention, www.ccsbt.org.
59 Ibid.
60 Fisheries Agency, *Waga kuni no suisan gaikō ni tsuite*, 9.
61 Ibid., 9.
62 Bergin and Haward, *Japan's Tuna Fishing Industry*, 146.
63 Polacheck, 'Experimental Catches and the Precautionary Approach', 285; and Anthony Bergin and Marcus Haward, 'The Political Economy of Japanese Distant Water Tuna Fisheries', *Marine Policy* vol. 25 (March 2001): 98.
64 Although North Atlantic bluefin tuna are a distinct stock from the SBT, the precedent that this might have created as well as the economic ramifications of registering and controlling the trade in all tuna species would have been very damaging to Japan.
65 Polacheck, 'Experimental Catches and the Precautionary Approach', 285.
66 Ibid., 285–86. In 1998, scientists from Australia and New Zealand estimated a low probability that SBT stocks would recover (< 14 percent), while Japanese scientists estimate a relatively high probability of recovery (76–87 percent). Results from Australian and New Zealand estimates also gave a greater than 50 percent chance that the parental biomass would continue to decline under 1998 quotas. The Japan Fisheries Association has reprinted the publications of a Japanese fisheries scientists in their monthly newsletter, *Isaribi*, which claims that tuna stocks are abundant worldwide and current estimates abroad consistently under-report population levels even in the South Pacific. Presumably, this is the position of Japanese tuna fishermen who lobby the Japanese government and offer statistical support to Japanese government delegations at the CCSBT.

67 Bergin and Haward, 'The Political Economy of Japanese Distant Water Tuna Fisheries', 98.
68 International Tribunal for the Law of the Sea, Award for the Southern Tuna Bluefin Case, 4 August 2000 (New Zealand vs. Japan; Australia vs. Japan), www.oceanlaw.net/cases/tuna2a.htm.
69 Ibid.
70 Ibid.
71 Ibid.
72 Ibid.
73 Ibid.
74 CCSBT website, www.ccsbt.org/docs/management.html.
75 Ibid.
76 Ibid.

Bibliography

Bergin, Anthony and Marcus Haward. *Japan's Tuna Fishing Industry: A Setting Sun or a New Dawn?* New York: Nova Science Publishers, 1996.

Doulman, David. *Fishing For Tuna: The Operations of Distant Water Fleets in the Pacific Islands Region.* Honolulu: East West Center, 1986.

Japan Fisheries Agency. *Waga kuni no suisan gaikō ni tsuite* (The Fisheries Diplomacy of Japan). Tokyo: Fisheries Agency, 2003.

Kang, Joon-Suk. 'The United Nations Convention on the Law of the Sea and Fishery Relations between Korea, Japan and China', *Marine Policy* vol. 27 (March 2003): 111–24.

Kim, Sun Pyo. 'The UN Convention on the Law of the Sea and New Fisheries Agreements in North East Asia', *Marine Policy* vol. 27 (March 2003): 97–109.

Marsh, James Barney, ed. *Resources and Environment in Asia's Marine Sector.* New York: Taylor and Francis, 1992.

Matsuda, Yoshiaki. 'Postwar Development and Expansion of Japan's Tuna Fishery', in *Tuna Issues and Perspectives in the Pacific Islands Region*, David Doulman, ed. Honolulu: East West Center, 1987.

Ministry of International Trade and Industry (MITI). *Keizai kyōryoku no genjō to mondaiten, Heisei 12-do* (The Conditions and Problems of Economic Cooperation). Tokyo: Keizai sangyō-shō, 2000.

Tarte, Sandra. *Japan's Aid Diplomacy and the Pacific Islands.* Pacific Policy Paper. Canberra: Australia National University, 1998.

Zou Keyuan. 'Sino-Japanese Fishery Management in the East China Sea', *Marine Policy* vol. 27 (March 2003): 125–42.

8 Epistemic norm formation and Japanese whaling policy

This chapter investigates Japanese whaling policy for the past several decades and tries to offer a distinctly political explanation for Japan's pro-whaling stance at the International Whaling Commission (IWC). While cultural and historic factors are important considerations, Japanese officials are also motivated by political goals whereby whaling and fisheries management is perceived as more than a mere resource allocation question, but as a food security issue as well. Japanese officials have been attempting to counteract what they see as a foreboding trend of growing environmental protectionism worldwide, not just in the field of whaling but in other fisheries as well, which threatens Japan's goal of food security. Whaling, as such, is a symbolic, precedent-setting case that will have repercussions for ocean management and allocation in international forums outside the IWC as well.

Whaling is a contentious international issue with important consequences for the bilateral relationship between Japan and the anti-whaling countries, including the United States, the United Kingdom, Australia and many other nations that are otherwise allies in other areas of economics and politics. This chapter will explore the reasons Japan continues to pursue a pro-whaling policy in spite of the grave risk of international opprobrium and economic sanctions faced if it resumes commercial whaling. This is especially perplexing in view of the relatively minor political influence of whaling communities in Japan and the marginal economic importance of the whaling catch.

As a single, compact issue that is widely familiar, Japan's whaling policy merits special consideration since it encapsulates the major themes of Japan's larger international fisheries policy. As with other kinds of fisheries, Japanese whaling faced increasing restrictions that were applied at the international level over the course of the post-war period. Motivated by a domestic desire to retain self-sufficiency in fisheries production, Japanese officials have responded to these restrictions with a combination of means that seek to reopen whaling upon the basis of scientific management and sustainable use, which would form the core of new rules governing international whaling.

Japan's contentious whaling stance is in part a fight for the rural economy, indigenous rights and the preservation of traditions, as is so frequently, if sometimes unconvincingly, asserted in the IWC. It will also be seen that whaling constitutes an important characteristic in the Japanese national consciousness.

Although these factors form a significant part of Japan's pro-whaling policy, Japanese policy is also premised on its food security strategy, which defends the fundamental right of access to all fisheries. In the case of whaling, which the Japanese assert is a fishery resource and not an environmental problem, this takes the form of a battle over management principles – the principle of 'sustainable-use' versus the principle of 'preservation' of a threatened resource. In short, Japan supports the scientific, rational use of resources in opposition to the protectionist encroachments and prohibitive moratoriums advocated by Western environmentalists. This issue hits at the very heart of Japan's longstanding quest for food security. It occurs at a time when Japan feels threatened by worldwide calls for greater environmental protection of fish stocks in the face of mounting evidence that worldwide fisheries resources are declining. This is especially poignant when Japan's consumption of fish is at record highs.

This chapter will start by examining the history of Japanese whaling to identify the key players and the conditions of the industry itself. This will be followed by an analysis of Japan's pro-whaling policy and its attempts to counteract what it sees as a foreboding trend of growing environmental protectionism worldwide, not just in the field of whaling but in other fisheries as well. Japanese officials wish to defend whaling's contribution to its national culture and are concerned that an expanding environmental movement threatens Japan's goal of food security.

The Japanese whaling industry

Japanese whaling techniques can be classified into two main types: coastal whaling and much larger-scale pelagic whaling. This chapter is primarily concerned with the second of these classifications.

It is worth making a few brief comments about coastal whaling since it is often this fishery that is featured by Japanese delegates when trying to create a palatable image for Japanese whaling in international forums. This fishery is largely subsistent and community-based, making it quite distinct from the large-scale, commercial operations of distant-water pelagic whaling. As the name implies, both types take place in Japanese coastal waters and they largely employ two capture techniques: harpooning and nets.[1] The number of whales taken by coastal whaling remained consistent at around 1,000 per year from 1957 to 1967, but has declined subsequently, making it no more than an artisanal industry.

Pelagic whaling, in contrast, is conducted by sizable fishing companies and whaling ventures are organized on an industrial scale. Only two of Japan's largest pre-war fishing and whaling companies, Taiyō and Nissui, were authorized to conduct whaling operations in the Antarctic, and even though new entrants were allowed after 1954 these two companies hereafter dominated the pelagic whaling industry.[2] Both of these companies are primarily fishing companies and whaling only constituted a small share of their combined industry. Although relatively small, their link to a much larger fishing industry gave these whaling companies some clout in Japanese politics, which may explain their being given exclusive access for so long.[3]

Early foundations of the post-war whaling industry

As was outlined in Chapter 2, the Japanese faced a food crisis of immense proportions immediately after the war, with the economy in shambles and much infrastructure destroyed by bombing. Soon after taking control in August 1945, the Supreme Command of the Allied Powers sought to secure a stable food supply for Japan, and the Natural Resources Section (NRS) set forth to solving the question of how Japan could overcome the problem of inherent resource scarcity, or the *shigen mondai*, and create a long-term solution to the food problem compounded by Japan's post-war isolation. Japanese economic planners worked with the Supreme Commander of the Allied Powers (SCAP) administration to devise strategies that could address short-term food scarcity and longer-term food security for the nation. Planners targeted fisheries (including whaling) as an industry worthy of speedy renewal since it could provide Japan with a much needed protein supply while also helping to foster the shipbuilding industry, despite such proposals provoking strong opposition from other Allied nations in SCAP.

The whaling effort rapidly expanded and Antarctic whaling ventures soon provided enough food to meet 47 percent of the per capita intake of animal protein by 1947.[4] It was also during this era that whale meat was distributed through a lunch box programme at schools throughout Japan and during which many older Japanese often fondly recollect their first taste of whale meat. In a country facing a severe food crisis, the Japanese press extensively covered Antarctic whaling hunts and the Japanese public saw these whalers as nothing less than saviours and heroes of the nation.[5]

Japanese pelagic whaling continued to expand at a remarkable rate in the post-war period. Whaling operations moved into the North Pacific in 1952 once the peace treaty was concluded and full sovereignty returned to Japan.[6] By 1958, Japan became the world's largest whaling nation in the Antarctic, whose catches surpassed those of Norway, Britain and the Soviet Union – a remarkable achievement given the extent of devastation the industry suffered just over a decade before.[7] Japan reached its all-time production peak in 1962 with a catch of over 300,000 tons of oil and meat, and was able to increase its catches despite the introduction of new restrictions, through purchases of foreign fleets and quotas.[8]

Japanese whaling leading up to the moratorium

Some mention should be made of the distinctive nature of the Japanese whaling industry. Whereas whaling was conducted in Western nations primarily for oil, this was not the case with the Japanese. Although whale oil was a valuable source of foreign currency in the 1930s and immediately after the war, the primary product of Japanese whaling ventures was meat. As was noted above, the revival of Antarctic whaling was justified on the basis of providing protein for a nation short on food resources. This continued to be the case, however, even well after the Occupation ended in 1952, with whale meat constituting 23 percent of animal protein consumed even as late as 1964.[9] The Japanese are also well known for using nearly every part of the whale for products as varied as cosmetics, *bunraku* puppets, jewellery and *koto* plectrums.

As the economy expanded after the war, however, consumer desires and preferences began to diversify, with major implications for the whaling industry. Other sources of meat became widely available once Japan surmounted its food crisis in the late 1940s, and as the price of beef, chicken and pork dropped with their increasing supply, Japanese consumers began to choose such livestock over whale meat.[10] Taiyō and Nissui were forced to rationalize their whaling operations as the market slumped in the 1960s and 1970s, since vessel reductions, cost cutting and other measures proved ineffective.

The whaling industry underwent a major restructuring in 1976 when the whaling sections of Taiyō, Nissui and Kyōkyo (Japan's third largest whaling company) were merged into a new company, Nihon Kyōdo Hogei.[11] When Japan started its research whaling programme after an international moratorium on whaling went into effect in the 1986/87 season, Nihon Kyōdo Hogei was disbanded and Kyōdō Senpaku was created in its place, hiring gunners and seamen, and procuring ships from the newly dissolved company.[12] Kyōdō Senpaku now operates as a not-for-profit chartered company under the aegis of the Institute of Cetacean Research (ICR). The Ministry of Agriculture, Forestry and Fisheries (MAFF), in turn, financially supports the ICR through subsidies and provides the mandate for their research programme, which is supposed to focus on the biological and socio-economic aspects of whaling.[13] More will be said of Kyōdō Senpaku and the ICR later.

Prior to the moratorium, whaling operations were conducted under the administrative guidance of the Fisheries Agency. These guidelines, including seasons, gear types and whaling quotas were subject to the controls established by the IWC. The Fisheries Agency and the Ministry of Foreign Affairs (MOFA) form the two core groups at the centre of whaling policy formation in Japan and both participate in annual IWC meetings. In contrast to other IWC members, however, an industry leader, rather than a senior fisheries regulator, regularly serves as Japan's IWC commissioner.[14] This is a good illustration of the extent of influence that the whaling industry exercises over whaling policy formulation. In short, the Fisheries Agency, MOFA and various industry groups, such as the Japan Whaling Association, work together to determine Japanese foreign policy with respect to whaling.[15]

Debate regarding scientific management within the IWC, 1946–82

An examination of IWC practices over the last five decades will reveal how the commission once served as an international body that supervised the exploitation of whaling stocks on behalf of whaling industries, but this function was inverted so that it now oversees their protection and preservation. During this time, an epistemological and ethical conflict regarding the use and management of whales emerged among whaling industrialists, cetologists and environmentalists which continues to this day. Japanese policy has been formulated largely as a response to the IWC's shift in design and, more importantly, as a means to counteract growing environmentalist concerns over other fisheries and environmental resources.

The IWC was established under the International Convention for the Regulation of Whaling (ICRW), which was signed in Cambridge, UK, in 1946. It comprised the major whaling nations of the world, which at that time included the United States, Great Britain, Norway and Iceland.[16] Japan was conspicuously absent at the foundation due to its status as an occupied nation, but joined the organization in 1951 as one of its first acts as a sovereign nation.

In its early days, the IWC's main responsibility was not the preservation of whales but the efficient allocation of whale stocks or the so-called orderly development of the industry.[17] This was done through the use of the blue whale unit (BWU) as a way to determine an overall catch quota. It was a measurement of the amount of oil one blue whale could yield: equivalent to two fin whales, two and a half humpback whales and six sei whales.[18] The BWU was not based on the relative abundance of whale stocks, but primarily upon industry needs. Since the IWC was mainly concerned with preventing the overproduction of whale oil as a means to stabilize world prices, the BWU served as a method to manage the whaling industry.[19]

In this sense, the BWU facilitated a kind of whale oil cartel. While the BWU helped the orderly development of the whaling industry, it proved disastrous as a means for managing a living resource facing decline. It did not make a distinction among the various whale species open for hunting, nor did it set levels based upon estimated population size or sustainable rates of capture. The BWU as a management principle was called into question as some great whale stocks began to face severe depletion and, in some cases, threat of extinction in the 1960s.

Although the ICRW specifically stated that decisions must be based on 'scientific' evidence, whaling science in the early phase of the IWC was not very well developed and the organization of scientific work was incomplete.[20] Until the late 1950s, 'cetologists lacked not only the agreed models that would produce quantifiable estimates of safe catch levels but also the stock size data needed to make conclusions from such models persuasive to regulators or whalers'.[21] One study into the problem of whaling science, which also hints at the problem of fisheries science in general, stated the quandary thus: 'It is exceptionally difficult to obtain even the most basic biological information ... because these species are totally aquatic ... Consequently it is very difficult to determine either the number of whales ... or changes in a stock size.'[22]

Moreover, whaling industry managers were able to exploit the uncertainty of fisheries science in order to resist any attempts to reduce quotas. Not only did industry managers have sufficient influence with enough governments to retain their influence over IWC decision making throughout the 1950s and 1960s, but the consistently high quota assessments encouraged over-investment in the whaling industry as each firm tried to catch as many whales as possible before the season ended.[23] Even as evidence emerged in the 1960s that some stocks of whales were being dangerously depleted, 'it was sufficiently weak and contestable that problems of recouping investment and protecting national industries against foreign rivals loomed larger in many participants' minds'.[24]

Thus the IWC in the early phases of existence was unable to strike a balance between the long-term objective to sustain the 'proper conservation of whale stocks' and the whaling industries' short-term motives to maintain the 'orderly development of the industry'. In order to strengthen the role of science in whale management and to counter the growing influence of industrial interests in the IWC, the United Kingdom proposed in 1960 to set up a special committee of three scientists, the so-called Committee of Three (C3) or the Three Wise Men, to offer outside advice and make recommendations to the Scientific Committee based on population dynamics.[25] In 1963, they submitted their report which recommended a total ban on catches of blue whales and humpbacks along with a significant reduction in overall quotas, and, most importantly of all, proposed in conjunction with the Scientific Committee that the BWU be abandoned as a management tool and that quotas should be established species by species.[26]

While the ban was adopted with respect to blue whale stocks, the proposed reduction in quotas was delayed for five years due to resistance by commissioners from Japan and the Soviet Union.[27] Reductions were finally introduced when whaling nations were unable to fill their maximum BWU quotas. The advice regarding the BWU management system was not adopted, however, and the BWU continued to be used until it was ultimately phased out in 1975, 12 years after the C3 recommendation.[28]

The BWU was finally replaced as a management tool with the introduction of a new model called the New Management Procedure (NMP) which, for the first time, established quotas for individual species of whales. Initially proposed by Australia, with strong backing from the United States, the NMP was adopted in 1974 and put into effect in the 1975/76 whaling season.[29] The NMP replaced the BWU as a management procedure, and established rules to ban whaling of over-exploited stocks while permitting catches of healthy whale populations based on a species-by-species quota. The NMP included estimates of natural mortality, pregnancy rates and the age at sexual maturity of whale species to determine catch limits derived from the maximum sustainable yield (MSY) model.[30]

Adoption of the new management procedure was a victory for conservationist cetologists over whaling industry managers who had stymied prior attempts to establish lower quotas based on a species-by-species management approach. It raised the profile of the Scientific Committee since the NMP was a management system that required greater scientific data and more accurate whale population dynamics, while also raising the amount of scientific argumentation that went into decision making.[31] The Scientific Committee used the NMP to create a more open decision-making process whereby scientific opinion was sought through publications and hearings, so that while it did not 'abolish political deals ... it made their use more obvious'.[32]

The adoption of the NMP, however, was not an unqualified victory for scientific management. The new model required precise population data and reliable models of population dynamics in order to develop accurate estimates of sustainable catches, but whaling science was plagued with high degrees of probable error and could not provide such accuracy at this time, resulting in a wide range of estimates.[33] Again, whaling industry managers and cetologists were at loggerheads regarding desirable quota limits. Before the dispute was resolved conclusively, however, another movement had taken over the IWC with the ultimate result of side-tracking both whaling interests and conservationist cetologists: the environmentalists' push to ban commercial whaling altogether.

Even before the adoption of the NMP, an increasingly influential environmentalist trend, originally based in the United States and the United Kingdom, was gaining momentum worldwide. Although the movement was concerned with the overall human impact on the environment in general, it became particularly concerned with the depletion of great whales and adopted the whale as a potent symbol for protecting the environment. As one study into the environmental movement concluded, 'saving the whale is for millions of people a crucial test of their political ability to halt environmental destruction'.[34] In response to the perceived threat of whale population extinction, protectionist-minded environmentalists pushed for a moratorium on all commercial whaling and lobbied their home governments through rallies and media appeals to push forward their ideas.[35]

In response to the rising popular support of the environmentalist agenda, the US government took the lead in several environmental issues, including whaling, in the early 1970s. It is important to note, however, that while environmentalists helped push US policy in the conservationist direction, they rarely fully determined it, as many Japanese officials believe.[36] At the first international conference on environmental issues in 1972, the United Nations Conference on the Human Environment (also known as the Stockholm Summit), the US delegation submitted a proposal calling for a moratorium on all endangered whale species. It cited the depletion of some great whale stocks, including the eight species of whales it placed on its own Endangered

Species List, and general mismanagement within the IWC itself as the justification for the ban.[37] The proposal was designed to stand for ten years while the IWC 'got its house in order', after which the moratorium would be reviewed.[38] Their proposal was supported at the United Nations conference by a 53:0 margin, with twelve abstentions, including Japan, which opposed the recommendation and labelled the resolution 'dramatic' and 'emotional'.[39]

A similar proposal in the IWC, however, subsequently failed to achieve the three-quarters majority required for implementation, where the Scientific Committee said that 'a blanket moratorium could not be justified scientifically'.[40] Japan succeeded in warding off attempts in the 1970s to impose a moratorium largely through winning important support of like-minded nations at the IWC. In the 1960s and 1970s, various fisheries-related purchases and investments helped the Japanese industry forge a set of transnational links with firms in other countries. This network became important during the intense competition for IWC votes in the 1976–82 period, since it gave the Japanese delegation local allies in a number of other IWC member countries.[41]

The next decade witnessed a pitched battle between pro-whaling and anti-whaling nations. Anti-whaling groups fought hard to change the IWC from an exploitative system to a preservationist one. The United States and influential environmental groups such as Greenpeace employed a tactic that essentially changed the membership, and hence the voting patterns, within the commission itself. In particular, they lobbied non-whaling nations to join the IWC and were successful to the extent that the membership in the IWC increased from 14 to 33 nations by 1982, with non-whaling nations eventually outnumbering whaling nations.[42]

With this new majority, the IWC General Council passed a measure in 1982 calling for a zero quota – which eventually became known as the 'moratorium' – on all species of whales irrespective of their stock status. This resolution passed despite the objections of the Scientific Committee. The moratorium would take effect in the 1986/87 whaling season and would be reviewed under a new management scheme by 1990.[43]

Japanese policy in the post-moratorium era

Japan strongly opposed the whaling moratorium. When Japan, Norway and Iceland threatened to exempt themselves from the provision, a right guaranteed under Article V, paragraph 3 of the ICRW, the United States warned that it might apply the Pelly and Packwood-Magnuson Amendments against them. Although Norway went ahead with the Article V exemption, Japan decided to comply with the moratorium when their fishing interests (mostly king crab and salmon) off the west coast of Alaska were threatened.[44]

The IWC hereafter changed in its function and practice. Whereas it used to be an organization largely dominated by whaling industry interests, it was later transformed into a conservationist organization that sought to protect whales. In its early stages, the IWC failed to incorporate science at the heart of management decisions in a timely manner and the newly adopted NMP was undermined by the industry's subsequent bickering over the uncertainty of scientific models and the allocation of species-specific quotas. Thus efforts to adapt the organization to a conservationist but pro-consumptionist management method failed. This bolstered environmentalists to abandon any efforts to compromise with industry interests in creating a scientifically managed whaling regime and, instead, to seek a total ban on all whaling. The adoption of the whaling moratorium signified the rising prominence of protectionist interests, whereby non-whaling nations and environmental organizations worked to redefine the aims of the IWC from a pro-whaling organization to one that would attempt to end commercial whaling altogether.

Japan has responded to the blanket moratorium in several ways. The moratorium was subject to full review in 1990 and although Japan had complied with the zero quota ruling of the IWC, it made every effort possible to reverse the ban during the 1990 IWC annual meeting in Noordwijk, Netherlands. These efforts failed, however, since anti-whaling countries still had a controlling majority in the IWC and were not amenable to seeing their earlier gains reversed. Since 1990, the Japanese government has been persistent in its efforts to have the moratorium lifted for whale stocks that have returned to numbers robust enough to sustain a whale hunt.

In particular, Japanese officials have consistently argued for a resumed hunt of minke whales, the one whale species that has returned to healthy numbers in recent years. Aware that a proposed revival of large-scale pelagic whaling in the Antarctic would not engender much support abroad, Japanese officials have carefully suggested a resumption of small-type coastal whaling of minke whales in order to support ailing whaling communities in rural Japan economically. Such resolutions have been submitted every year since 2002, but have not yet successfully won ratification by the IWC.

Japanese policy makers are also keen to remove the earlier-cited basis for the moratorium by rectifying the insufficiency of scientific knowledge on whale populations and to develop a new management regime. Consequently, the Fisheries Agency has initiated a scientific programme to study whales in addition to developing and promoting a new scientific management regime, the Revised Management System (RMS). MOFA and the Japanese Whaling Association are both strong supporters of these efforts.[45]

It should also be noted that, shortly before this book went to print, Japan lost a landmark case in the International Court of Justice with regard to its scientific whaling program in the Antarctic Ocean.[46] It has henceforth been ordered to desist from such whaling. Although the Government of Japan announced that it would comply with the ruling, it also announced that it would start preparations for a resumed expedition in the North Pacific as of Autumn 2014 within only a few months of the ICJ decision. It is too early to say how events might develop, but surely an understanding of the background of scientific whaling and its official rationale merits special consideration.

In order to improve the scientific understanding of whale species and their population dynamics, Japan has administered a research whaling programme under the auspices of the IWC's Article 8, paragraph 1 since 1987. These studies are authorized under MAFF and conducted by the ICR. Since 1989, Japan has annually caught around 400 minke whales as part of its lethal-methods scientific whaling programme focusing on the Antarctic.[47] Since 1994, research fleets have also been dispatched to the North Pacific to test samples of minke whale populations, along with an additional catch of about 50 Bryde's and 10 sperm whales.[48]

According to the ICR, Japanese research whaling is conducted to acquire knowledge about population structure and feeding habits of targeted whales.[49] Although some research involves sighting, skin sampling and other non-lethal techniques, most studies require that whales be killed in order to collect instructive statistics. This has proven very controversial and many people, scientists and environmentalists alike, have questioned the validity of lethal methods, suggesting that population and health studies may be best achieved through simple observation, skin sampling or faecal records.[50]

Nonetheless, the Japanese argue that lethal methods are the most effective in determining the feeding patterns and age of whales. By capturing whales, scientists can establish the age of the whale by counting the rings formed on its inner earplugs (naturally, this can only be done on a dead whale), establish the gender of the animal, as well as investigate the stomach contents of the animal to model its feeding habits – all data necessary to create accurate population estimates.[51] Chemical and DNA analyses are also done to test the toxicity levels of the Antarctic and North Pacific, and to identify sub-groupings within a local population.

Other observers have noted that scientific whaling is not only designed to acquire new understanding about the whales themselves, but to demonstrate that targeted whales are not endangered. In other words, samples are taken from the population to demonstrate that the whale species in question can sustain a full-scale hunt, as in the case of the minke whales. These methods are somewhat dubious, however, since no one can be sure whether a particular species is endangered in advance of the sampling.[52]

According to IWC rules, the ICR is entitled to market the products of this hunt, even though this has raised the ire of environmentalists worldwide. This is done so as to minimize the amount of waste from such studies. These sales are able to fetch about US$35 million a year in the local market and the products normally find their way to high-priced restaurants in Tokyo. The research programme itself, however, costs about US$40 million a year, which requires an annual government subsidy of US$5 million to offset this short-fall.[53] One *New York Times* article describes the subsidy thus: 'The government uses taxpayer money to persuade taxpayers to eat whales hunted with taxpayer money.'[54] While there is little evidence showing a direct connection between governmental subsidies and the marketing of whale meat, as this article suggests, the Fisheries Agency sees the scientific whaling programme's preservation of tradition as an important function.[55] Although the hunt is ostensibly for research purposes, the promotion of whale-eating and the pre-servation of harpooning and flensing techniques are no doubt a salient component of the continued whale hunt in Japan.

To accompany the programme for scientific research, Japanese officials and scientists have also been instrumental in developing new management rules for whaling. Japanese delegates at the IWC propose to administer a renewed whale hunt under the RMS, a management system that conservatively controls for scientific error and uncertainty.

Central to RMS guidelines is a quota calculation tool called the Revised Management Procedure (RMP), which uses mathematical models to calculate safe catch quotas for stocks that are plentiful and for which there is a great deal of uncertainty about population numbers and dynamics. Using estimates of current abundance of whale stocks taken at regular intervals coupled with knowledge of past and present catches, the RMP employs an algorithm to determine very conservative quotas that aim to be as stable as possible, to exempt stocks estimated to be below 54 percent carrying capacity, and to achieve the highest possible continuing yield from the stock.[56] The Scientific Com-mittee has unanimously recommended the RMP to the commission, and all the scientific aspects of the work were adopted by the commission in 1994. Its actual implementation in whale management, however, has been delayed until the moratorium is lifted.[57] As Ray Gambell, former secretary to the IWC, noted recently, 'The RMP is robust to a range of factors, including under-estimation of historic catches by up to 50%, variations over time in carrying capacity and recruitment, environmental degradations and a wide range of uncertainty, including stock units and differing population dynamics'.[58]

The RMS, on the other hand, establishes an inspection and observation scheme that must be in place before any reversal of the moratorium can be considered. Currently under discussion in the IWC and fully endorsed by the IWC's Scientific Committee, the RMS proposes to establish a system of national inspectors and international observers to be present on whaling fleets, and full monitoring of vessels to verify the types of whales caught.[59] There have even been proposals to use satellite tracking as a means to moni-tor whaling operations and to assist in estimations of population sizes and migration routes.[60] In general, the RMS aims to renew limited commercial whaling under a very conservative management scheme that errs in favour of sustaining whale populations.

Another condition necessary for the reversal of the moratorium and the implementation of the RMS is the presence of a whale stock that has recov-ered to a level healthy enough to support a sustained hunt. Japan argues that this is just the case with various stocks of minke whales, the most abundant of which are found in the Antarctic. Official Japanese estimates conclude that as many as four times the number of whales that Japan currently takes (totalling 1,600 whales per year) could be caught safely for a century.[61] A commonly accepted figure among whale experts in the 1990s estimated that there were nearly 760,000 minke whales in the Antarctic, more than enough to sustain a limited catch.

The IWC's Scientific Committee, however, has since been in a deadlock over estimates as some scientists believe the number to be appreciably lower than earlier figures. One of the sources of controversy derives from recent revelations that Soviet whalers had routinely ignored IWC quotas in the past while submitting drastically lower catch statistics to the IWC's Scientific Committee.[621] Scientists thus fear that current numbers of all whale populations are much lower than originally thought. Nonetheless, Japanese officials still contend that minke whales are so prolific in the Southern Hemisphere that their eating habits are depleting stocks of fish in direct competition with fishermen and to the detriment of other whale species.[63] One official even likened them to 'cockroaches' of the oceans due to their abundance.[64]

In conjunction with the scientific and management programmes designed to revive the whale hunt, Japanese officials have also used political means in an attempt to regain a voting majority within the IWC itself. At present, the IWC's thirty-eight voting members are nearly split evenly between those supporting and opposing the whaling moratorium.[65] The vote, however, needs a three-quarters majority to be passed. Nonetheless, this represents a significant reversal of the three-quarters majority that the anti-whaling lobby held in the early 1980s.

This reversal has largely been achieved through Japan's recruitment of smaller member nations in the IWC as well as the signing up of new member states sympathetic to Japan's cause, a tactic that closely mirrors those used by the anti-whaling lobby in the late 1970s. Some observers allege that Japan uses its official development assistance (ODA) policy to 'buy' votes and influence the decisions of smaller member states – namely, Grenada, Antigua and Barbuda, Dominica, St Lucia, St Vincent and the Grenadines, the Solomon Islands and Guinea – by either offering new grants for development projects or threatening the withdrawal of support for others.[66] Greenpeace has also accused Japan of paying the rather expensive IWC membership fees on behalf of these nations – an ironic allegation since environmental groups used the very same tactic to boost their position during the moratorium vote in 1982.[67] Japanese officials staunchly deny these allegations and instead suggest that Japan has won such support due to its scientific arguments in the IWC and because of these countries' commitment to international law and institutions.[68]

The goals of Japanese whaling policy

Japan conducts research whaling in an effort to provide the scientific knowledge and evidence of whaling stock renewal needed to lift the whaling moratorium. Even though scientific whaling is permitted by the IWC, there is still vocal pressure from the United States, the United Kingdom, Australia and New Zealand to discontinue whaling altogether, scientific or otherwise. Much of this pressure is generated by the relatively powerful environmental lobbies in these countries, which argue that any type of whale killing is morally wrong, whether for scientific, cultural or economic reasons.[69] It is often argued by these groups that whales are special creatures deserving superior status to other types of 'food animals' by virtue of intelligence, size and mammalian nature. They question why Japan cannot adapt, much like other former whaling nations, to a non-whaling existence.

Pro-whaling Japanese are quick to debunk such arguments as the imposition of foreign norms upon a country whose whaling culture runs deep into the history of the nation. Much ink has been spilled detailing the distinctive whaling culture of Japan, the very existence of which is threatened by the international moratorium.[70] Officials have argued that the Japanese hunt whale primarily for meat, and not oil as in the West, and thus whales have earned a special place in the food culture of a nation dependent upon marine resources.[71] Some Japanese officials have even described eating whale meat as part of the genetic make-up of Japanese – another feature that separates the Japanese from their Western counterparts.[72]

Moreover, local fishing communities that have been established upon the whaling hunt have few alternative economic activities or fisheries to support those laid-off from work as a result of the moratorium. For these rural communities the moratorium could spell the end of a traditional way of life.[73] The extreme among Japanese see the moral arguments made by foreign environmentalists as racially motivated and tantamount to 'cultural imperialism'.[74] Some Japanese officials have even attempted to get whaling in Japan classified as an 'aboriginal' fishery so as to get it permitted under IWC rules, although this has gained very little sympathy from IWC members. In any case, many Japanese feel that whaling forms an important part of their lifestyle, both in traditional rites and art as well as in culinary preferences, such that its disappearance would be seen as a tragic end to a unique form of Japanese culture.[75]

Finally, whaling occupies a special place in the hearts of the older generation within Japan. Many middle-aged and elderly Japanese fondly remember the taste of whale meat and wish that it would become available once more. To such people, the whaling ban represents nothing less than a travesty brought upon Japan from outside; some have even likened the destruction of this way of life to the destruction of traditions brought about after the Second World War. It is hard to deny that whaling features as an important part of the contemporary national character of the Japanese; for some, it even represents a nationalist issue.[76]

Despite the cultural significance of whaling to many Japanese, its economic importance is not so well defined. Although whaling provided an important industry to several smaller coastal towns, its economic contribution on a national scale was traditionally quite small. Whaling has been a marginal activity in Japan's overall fishing effort in the post-war period and its continuation was only guaranteed after successive mergers and rationalization plans adopted in 1976, as discussed before. Even before the imposition of the moratorium in 1987, the industry was declining in importance relative to other fisheries. The marginal importance of whaling to overall fisheries production can be seen in Figure 8.1.

Even pelagic whaling was primarily an activity of small, remote coastal villages of little significance to overall national politics. Whaling communities as a whole do not command many resources or electoral significance with which to capture the attention of politicians or bureaucrats. At the time of the imposition of the whaling moratorium in 1987, only 930 people were directly employed in whaling. Even a decade before the imposition of the moratorium, Nihon Kyōdo Hogei, formed as an amalgamation of Japan's largest whaling companies, directly employed only 1,500 people.[77]

Whale products have only dubious value to the local economy in terms of profit or benefit. Whale meat nowadays commands no more than a minority taste in Japan and the high prices fetched in some of Tokyo's restaurants have more to do with the price some middle-aged people are willing to pay for its nostalgic value than its widespread popularity or high quality. The high price is also a product of the extra relative demand that is created by a very limited supply. Indeed, surveys show that only 1 percent of Japanese eat whale meat more than once a month.[78]

Many Japanese officials interviewed have also stated that one of the primary reasons for the government's support for the resumption of whaling is that it has widespread popular support in Japan. A review of polls conducted over the last several years, however, is not so conclusive. For example, a government survey conducted in 2002 indicated that 70 percent of Japanese support the resumption of commercial whaling.[79] In a poll conducted the same year by the *Asahi Shinbun*, however, very different results could be seen. They

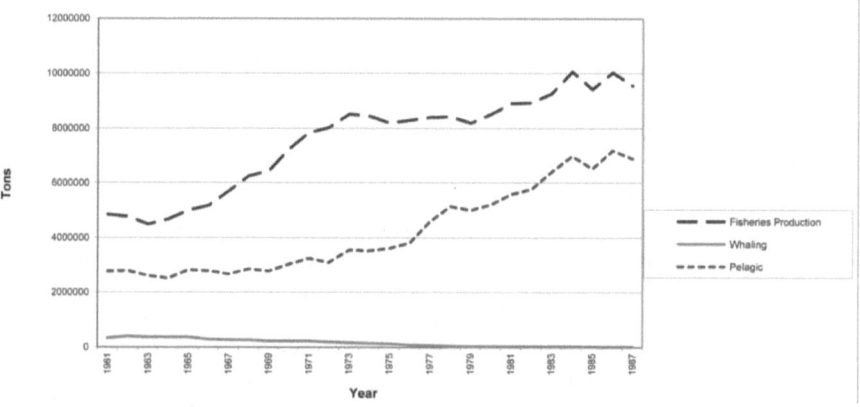

Figure 8.1 Japanese fisheries and whale production, 1961–87
Source: FAO, Fishstat Database, www.fao.org/fi/statist/FISOFT/FISHPLUS.asp; Arne Kalland and Brian Moeran, *Japanese Whaling: End of an Era?* London: Curzon Press, 1992, 199–200)

found that almost 60 percent of Japanese in their twenties *opposed* the resumption of commercial whaling and that only a mere 4 percent of all respondents ate whale meat 'sometimes'.[80] While it is uncertain how popular the resumption of commercial whaling is for the average Japanese, there is little doubt of how dear this issue is to the hearts of Japanese officials. It would also seem that there is a generational gap between older Japanese who support whaling and the younger generations who are against it.

In consideration of the relative economic insignificance of whaling and its remote position with respect to national politics, it is fair to ask why whaling features so prominently in Japanese foreign policy today. It is also worth considering why Japan does not withdraw from the IWC, much as Iceland did in 1991, if it creates so much trouble for their whaling efforts.

Some Fisheries Agency officials have argued that there may come a day when Japan withdraws from the IWC altogether if it proves intransigent or becomes 'hijacked' by environmentalists.[81] Such a position, however, is firmly denied by MOFA. Despite some hints and threats that it may pull out of the IWC, Japan has elected to stay within the international organization for a number of reasons. First of all, the IWC is recognized as the only official international body that regulates whaling. Withdrawing from the IWC would be tantamount to Japan turning its back on international diplomacy and international law, both of which it strongly upholds in its foreign policy. One MOFA official went as far as to suggest that such a withdrawal would compare to Japan's exit from the League of Nations in 1931.[82] By approaching the IWC to authorize its whaling operations, Japan hopes to be able to legitimize future whaling operations under international law.

Another reason Japan elects to participate in contentious IWC discussions is because of the symbolic value of whaling as a precedent-setting issue. Japan is particularly concerned about the larger question of sustainable resource use in general, and a victory in the whaling debate would be a triumph for Japan's arguments regarding the use and distribution of increasingly scarce maritime resources. As Japan's former representative to the IWC, Yonezawa Kunio, once stated: 'For Japan to withdraw from the IWC might please extremists, but it would not necessarily help our concern for sustainable whaling or further our larger cause ... it encompasses much broader questions, among which are the fundamental human right to use natural resources responsibly.'[83] He is alluding to Japanese efforts to protect the notion of the scientific management of natural resources and, more specifically, valuable fisheries stocks in the North Pacific and elsewhere. Without this principle, Japan will be unable to argue for managed exploitation of resources as they become scarcer and may lose secure, equal access to more important marine resources as a result. The defeat of this principle would strengthen the move of preservationists to protect other fisheries when threatened and would compromise Japan's more general goal of food security.[84] Moronuki Hideki, an official at the Fisheries Agency, clearly identified the risk that environmentalist moratoriums would have on the future of whaling and fishing in

Japan: 'If the current ban on hunting whales is allowed to become permanent, activists may direct their efforts to restricting other types of fishing.'[85]

Both MOFA and the Fisheries Agency now promote 'sustainable use' since they are concerned that the preservation principle upon which the moratorium is premised may gain wider adherence globally. MOFA has sponsored several conferences in recent years that address 'sustainable use of marine resources' as the core element of the whaling debate.[86] The Fisheries Agency is also a strong advocate of sustainable use for whales. As the assistant director of the Far Seas Division declared, 'if we give up [sustainable use of whales] in the field of whaling this creates a bad precedent for Japan which has major fleets for fisheries other than whaling'.[87] The ICR declares that one of its chief aims is to promote the sustainable utilization of marine resources because:

> Problems surrounding Japanese fisheries have become increasingly complex in recent years in relation to the conservation and management of marine mammals, as exemplified by the IWC's adoption of a moratorium on all commercial whaling in 1982 and the Southern Ocean Whale Sanctuary in 1994, as well as by the tightening of regulatory measures on driftnet and other fishery technology ... More restrictive measures are expected to be imposed internationally upon fisheries, including high-seas fisheries.[88]

The former deputy director-general of the Fisheries Agency, Shima Kazuo, explains that anti-whaling campaigns have escalated into an anti-fishing movement and there exists a real risk that fisheries now deemed under threat might be protected, much like the whale, by sanctuaries or moratoriums: 'Those of us who uphold the principle of the rational uses of marine living resources must fight those against by renovating the concept of marine resource management, and correcting the direction of the present global drift.'[89]

For Japanese officials, the cornerstone of sustainable use is the scientific management of the resource in question. What they are combating is an increasingly widespread preservationist movement that does not share their faith in science to manage marine resources, but which instead calls for moratoriums, sanctuaries and other restrictive systems that have the principle of non-use at their core. Such institutions may also be termed prohibition regimes, and they have already been developed in the past to protect elephants, fight the drug trade, protect human rights and ban cod fishing in Eastern Canada. Ethan Nadelmann argues that 'global prohibition regimes almost certainly will play an increasingly prominent role ... in efforts to protect newly endangered species, to reduce pollution of the seas and the skies, and to conserve forests and other dimensions of the earth's natural resources'.[90] Japanese officials are concerned that this trend may become increasingly applied to valuable marine fisheries if evidence of overfishing or depletion comes to light.

Policy and science

Prohibition regimes have been applied in the past when a resource became alarmingly scarce due to either poor management or poor science. For example, the introduction of the whaling moratorium was premised on the poor state of management systems as well as an inadequate knowledge base for science, both of which led to the unexpected depletion of whale populations. Preservationists question the ability of fisheries biologists effectively to manage whale resources and often focus on the 'uncertainty of science' as a mandate to cease all exploitation altogether as the only means to allow an endangered renewable resource to recover.

Whaling scientists face the same management dilemma as other fisheries scientists: how does a scientist provide reliable population figures, and hence reliable catch estimates, when fish cannot be counted in the vast oceans? This dilemma arises from the debate about the exactitude of fisheries science in determining population levels and prescribing sustainable catch rates. In short, it focuses on the ability or inability of scientists effectively to manage fishing. As Michael Berril, a noted Canadian fisheries biologist, once observed, 'pity the fisheries biologist. They can't be certain, and their best estimates will likely be misunderstood, misused and perhaps abused'.[91] Many environmentalists and more cautious fisheries biologists are sceptical of the ability of science to provide fisheries managers with reliable data, even with newly developed mathematical models. Such people interpret uncertainty as a fiat to place severe restrictions on resource use, including moratoriums, as the only means to allow severely depleted renewable resources to recover.[92] They believe that there have been no substantial improvements in scientific techniques in recent years and are thus hesitant to re-open whaling or fisheries to renewed exploitation. Indeed, some environmentalists see the earlier mismanagement and near-extinction of some whale species as a symptom of fisheries science's inability to manage whale populations and catch levels at all.

There are even more parallels between whaling and other fisheries besides 'scientific uncertainty' that make the whaling case a possible precedent for other endangered fisheries. Much like the Stockholm declaration that preceded the whaling moratorium, environmentalists have been calling for the creation of a new international regime to preserve highly migratory stocks outside of national fisheries zones.[93] This was recently addressed at the World Development Summit held in Johannesburg in 2002, when an agreement to restore global fish stocks was one of the few issue areas agreed upon by international delegates.[94] This bears some of the features of the early environmentalist campaigns to 'save the whales' during the Stockholm Summit in 1972, a decade before the adoption of the whaling moratorium. Moreover, environmentalists are beginning to step up pressure to halt the most destructive kinds of high seas fisheries, as could be seen through the widely publicized case of 'dolphin-free' tuna branding.[95]

There is significant evidence of declining fish stocks in commercially valuable species in many oceans of the world, which has led to rising environmentalist calls to address the problem of overfishing. An overview of the state of exploitation of some of the major fish species worldwide reveals a similar story. According to the Food and Agriculture Organization's (FAO) *White Paper on World Fisheries* (2000), 25–27 percent of the major marine resources of the world remain unexploited or semi-exploited, 47–50 percent are fully exploited, 15–18 percent overexploited, and depleted resources and those on the way to recovery account for 9–10 percent.[96] Some notable cases of endangered fish stocks, as listed in the World Conservation Union (IUCN) Red List (2000), include: albacore and southern bluefin tuna, which are now listed as critically endangered; the Pacific bigeye tuna, as endangered; and northern bluefin tuna, as either endangered or critically endangered depending upon the region.[97]

Many new fisheries biology and population studies have suggested that the world fish catch, much like whaling in the 1970s, has been declining in recent years as a result of overfishing, and this has had a negative impact on ocean habitat and marine ecosystems. One such study by Reg Watson and Daniel Pauly caused shockwaves among marine biologists and fisheries managers by finding that despite world FAO statistics indicating a consistently rising world catch rate since 1955, world catch rates in fact have been declining since 1988.[98] The implication of this is that whereas marine biologists believed that fish populations were healthy due to rising catches (indicating more fish in the sea), there are actually fewer fish than previously thought. These misleading trends resulted from repeated falsification of fisheries statistics by the world's largest fish producer, the People's Republic of China, which has distorted overall estimations of fish populations in favour of a much higher count than actually exists.[99] Now that it is understood that the global catch is declining, then fish stocks must be declining too.

These findings were corroborated by a fisheries population study conducted by Ransom Myers and Boris Worm in 2003. Researching the effects of fishing upon populations of predatory fish (consisting of mostly commercial species), they found that since 1950, with the onset of industrialized fisheries, the fisheries resource base has been reduced to less than 10 percent: not just in some areas or just for some stocks but for entire communities of large fish species around the world.[100] Using the most current data and population models, this study provided the scientific evidence that confirmed the worst fears of marine biologists, who had hitherto made only vague and frequently over-optimistic guesses about the status of fish stocks. Years of overfishing in unregulated or under-regulated fisheries are now evidencing a toll upon commercial fish stocks in all oceans of the world. As a result, fishermen are targeting smaller, less commercially valuable fish that are relatively more abundant, in a process that has been aptly described as 'fishing down marine food webs'.[101]

Opinions on the status of fish stocks are not only changing in the scientific community. The public at large is becoming more interested in the ocean environment and the problems associated with declining fish populations and overfishing. Due to the widespread appeal of the BBC's documentary movie, *Deep Blue*, many more people have become more aware of the ocean and its ecosystems. Although, regrettably, no explicit reference was made to fishing, the documentary provided a stunning underwater view of a realm of Earth few people have seen firsthand. Charles Clover has recently published a popular book entitled *End of the Line* which, for the first time in paperback format, makes an intelligent condemnation of the world's most wasteful fisheries and explains the consequences poor fisheries management has upon the future of ocean ecosystems.[102] As Clover states in the introduction of his book, 'the perception-changing moment for the oceans has arrived. It comes from the realization that in a single human lifetime we have inflicted a crisis on the oceans greater than any yet caused by pollution ... there is no exaggeration in saying that overfishing is changing the world'.[103]

Contemporary scientific and public concern for overfishing and declining fish stocks mirrors the alarms environmentalists raised on whaling during the Stockholm Summit in 1972, which eventually led to widespread calls for a moratorium. Fisheries are growing into an issue with broader public interest than any time before and this will likely entail greater support for environmental non-governmental organizations' (NGOs) efforts to protect and conserve world fisheries that have, until now, been allowed to operate outside the public's view. There has already been evidence of this with the widespread acceptance of an NGO proposal for sustainable fisheries that was one of the few resolutions eventually supported by both states and environmental groups at the Johannesburg Earth Summit in 2003.

Many Japanese officials in both the Fisheries Agency and MOFA recognize that the future of Japanese fisheries is in some doubt due to declining productivity from the oceans and rapid global population growth.[104] Japanese officials are trying to plan for an era of greater scarcity while also renewing efforts to foster greater self-sufficiency. As one Fisheries Agency official states in his book, *Whales and the Japanese*:

> When we ponder an increasingly over-populated world, we must also look at the uneven distribution of food. Japan cannot continue simply relying on imported food. Can we afford as a country to be dependent on others, such as the United States or Australia, for our basic foods? Will we always have enough precious dollars to import what we need? It is the answers to these questions that should tell you why I firmly believe that we need to become more self-sufficient for reasons of our national health and at the most basic level, to guarantee the supply of food to our people.[105]

Mr Komatsu then goes on to say that opening the whaling industry would help alleviate the dangers of declining sources of food protein, but the same could be said of trying to prevent the closure of any other fisheries.

In its efforts to face a perceived shortage of fisheries products in the future, the prime minister's office set a target of 65 percent self-sufficiency in food sources to be achieved by 2010.[106] At present, food self-sufficiency in terms of calories in Japan is only 41 percent.[107] Some industry representatives said that an increase in self-sufficiency may be achieved if there was a significant revival of whaling effort.[108] One former whaler even talked of reviving Japan's custom of serving whale meat in elementary school lunchboxes as a way to deal with future protein supply shortages.[109]

Conclusion

Early mismanagement at the IWC and the hubris of industry leaders forestalled the adoption of scientific models that would have entailed stricter quotas to protect declining whale stocks. When such efforts to establish a scientifically managed system faltered due to internal disputes or were diluted to appease industry interests, another, less-appeasing movement led by environmental groups soon took over and imposed a comprehensive moratorium on whaling at the IWC.

Japan wishes to overturn the ban on whaling that has been in place since the 1996/97 whaling season. Japanese officials are trying to renew the IWC's faith in science so that a managed fisheries regime, a 'sustainable-use' regime, can be created, instead of the current trend of irreversible moratoriums. Japanese officials have sustained the argument that whaling can be conducted on a scientific basis and have been committed to acquiring population data to support the reopening of whaling for certain species, particularly the Minke whale, which Japanese scientists claim have a population large enough to sustain a limited catch. The motivation for Japanese whaling policy is partly to protect the culture and traditions of a nation that historically harvested whales. Japanese officials are also apprehensive that the IWC whaling ban may serve as an adverse precedent, whereby a similar system of irreversible moratoriums is applied to more valuable migratory fishing stocks, such as tuna or salmon, as they similarly become depleted due to excessive fishing pressure and inadequate management.

The whaling issue is important to the Japanese because it serves as a symbol of the kind of threat that unbridled conservationism may pose to all of Japan's future fisheries and, hence, to its food security. The shift of the IWC from an exploitative regime to a protectionist one reveals the extent to which international norms may be mobilized to capture political power for a specific environmental cause. It is feared that if other world fisheries achieve the notoriety and sympathy for protection that whales have, then they too may be closed, never to reopen, much as the whaling fishery to date.

Notes

1 More detailed descriptions of the history and operations of coastal whaling can be found in Arne Kalland and Brian Moeran, *Japanese Whaling: End of an Era?* London: Curzon Press, 1992, 84–88.
2 Kalland and Moeran, *Japanese Whaling*, 92.
3 For a detailed discussion of whaling industry links to politics, consult Yutaka Hirasawa, 'The Whaling Industry in Japan's Economy', *The Whaling Issue in U.S.-Japan Relations*, John R. Schmidhauser and George Totten III, eds. Boulder, CO: Westview Press, 1978, 82–114.
4 Masayuki Komatsu and Shigeko Misaki, *Whales and the Japanese.* Tokyo: Japan Whaling Association, 2003, 23. It is worth noting, however, that in terms of overall protein consumption, whale meat provided a mere 1 percent of intake (Harry Scheiber, *Inter-Allied Conflicts and International Law, 1945–53: The Occupation Command's Revival of Japanese Whaling and Marine Fisheries.* Taipei: Academia Sinica, 2001, 131). This was because very little animal meat was consumed relative to protein from vegetables such as soy beans.
5 Junichi Takahashi, 'Whaling Culture in Contemporary Japan', *The 1st Summit of Japanese Traditional Whaling Communities.* Tokyo: Institute of Cetacean Research, 2002, 63.
6 Kalland and Moeran, *Japanese Whaling*, 89.
7 Komatsu and Misaki, *Whales and the Japanese*, 66.
8 Kalland and Moeran, *Japanese Whaling*, 89–90.
9 Kalland and Moeran, *Japanese Whaling*, 90.
10 Junichi Takahashi, 'Whaling Culture in Contemporary Japan', 67; and Fisheries Agency, *Suisan Hakusho 1975* (Fisheries White Paper). Tokyo: Ministry of Agriculture, Forestry and Fisheries, 1976, 7.
11 Kalland and Moeran, *Japanese Whaling*, 93. Officials in the Japan Whaling Association suggested during an interview (18 February 2004) that the 1976 merger of the three largest whaling companies was due to a falling supply of whales for the domestic Japanese market. They blamed the falling catches upon increasingly restrictive quotas imposed upon the industry by the International Whaling Commission.
12 Interview with Ōsumi Seiji, Director General, Institute of Cetacean Research, 2 March 2004.
13 Interview with Ōsumi Seiji.
14 M.J. Peterson, 'Whalers, Cetologists, Environmentalists, and the International Management of Whaling', *International Organization* vol. 46 (Winter 1992): 171.
15 Interviews with officials at MOFA and the Fisheries Agency (February and March 2004).
16 The IWC is composed of commissioners who are nominated by member nations. A general assembly meeting is held every year at which commissioners discuss outstanding issues concerning the allocation of whale catches, scientific findings and whale management. The IWC is also composed of three committees: the Scientific, Technical and Finance/Administration Committees. The Scientific Committee advises on the state of whale stocks and suggests safe quota levels for whaling. The Technical Committee considers matters pertaining to regulation and can recommend policies on whaling seasons, gear types and the like, and has evolved into preliminary meetings of commissioners acting on behalf of the whole organization. The Finance and Administration Committee oversees the location and preparations for meetings as well as the operations of the relatively small secretariat, which consisted of only two people until 1974 and only a handful thereafter.
17 According to the IWC's Constitution, the Commission's principal function is to oversee the 'orderly development of the whaling industry' and to ensure the

'proper conservation of whale stocks' based on scientific findings (International Convention for the Regulation of Whaling, 'Preamble', Washington, DC, 2 December 1946, www.iwcoffice.org/Convention.htm). It is charged with the responsibility of managing specific whale stocks that include all species of baleen and toothed whales with the exception of small cetaceans, such as dolphins and porpoises. The IWC is the first multilateral body of its kind to adopt the requirement that conservation of a living marine resource be based on science.

18 Japan Whaling Association (JWA), www.whaling.jp/english/history.html#04.
19 Once the worldwide BWU catch quota was decided by the IWC, the 'Olympic' or 'first come first served' method was used to determine the allocation, which inadvertently resulted in intense competition among whaling nations over finite stocks (www.whaling.jp/english/history.html). Some degree of control, however, was exercised among IWC member states. Although the IWC did not have the authority to set national quotas prior to 1961, the major whaling states preferred to negotiate among themselves to determine allocations (J.N. Tonnessen and A.D. Johnsen, *The History of Modern Whaling*. Berkeley, CA: University of California Press, 1982, 596–97).
20 The IWC is not a binding organization and the commission does not have the power to set regulations that limit entry into fisheries, enforce quotas or any other such measures that would in any way dilute national control of interests. At best, the Commission can make recommendations and it is left up to the commissioners' vote or consent as to whether they are put into effect. As such, the national commissioners are the highest authority within the IWC. Even when the IWC makes a recommendation or suggests regulations, such as a total ban of commercial whaling, non-obliging member states have two principal options to avoid its application. First of all, they can legally exempt themselves from the rules by filing a formal objection under Article V, paragraph 3, which nullifies its application to the member state. (International Convention for the Regulation of Whaling, www.iwcoffice.org/Convention.htm). Second, they can withdraw from the Convention altogether, since there is no international law requiring the participation of whaling nations in the organization. The IWC's limited power to adopt regulations and lack of authority to enforce them has led one authority to comment that the Commission is 'a thing to be captured rather than an actor in its own right' (Peterson, 'Whalers, Cetologists, Environmentalists', 156).
21 J.A. Gulland, 'The Antarctic Treaty System as a Resource Management Mechanism', *The Antarctic Treaty Regime*, Gillian D. Triggs, ed. Cambridge: Cambridge University Press, 1987, 119.
22 J.E. Scarff, 'The International Management of Whales, Dolphins and Porpoises: An Interdisciplinary Assessment', *Ecology Law Quarterly* vol. 6 (1977): 333. The poor state of whaling science raised a great deal of uncertainty in the information and knowledge upon which the IWC had to make its decisions and quota recommendations. Although fisheries biologists during the 1950s generally agreed that quota levels were set too high, they were unable to quantify their advice and elected instead to wait for better population data and calculation techniques in the future. In doing so, they failed to provide the management guidelines that might have reduced quotas in a timely manner.
23 Peterson, 'Whalers, Cetologists, Environmentalists', 161.
24 Ibid., 158. Meanwhile, due to lower demand of whale oil and the substantial costs involved in sending fleets to the Antarctic, many industry managers found that Antarctic whaling ventures were becoming less viable to the extent that several nations abandoned whaling in the Antarctic altogether: the UK stopped whaling after 1963, the Netherlands and New Zealand after 1964 and Norway (in Antarctica) in 1968. The Japanese and the Soviets, however, had a substantial home market for whale meat that provided a lucrative alternative to cheaply priced

whale oil and were thus able to continue their operations throughout the 1960s and 1970s (George Small, *The Blue Whale*. New York: Columbia University Press, 1971, 39–42). In fact, even when catch limits were reduced, Japan continued to run profitable operations and even purchased quotas from Great Britain and Norway to expand its catch (Komatsu and Misaki, *Whales and the Japanese*, 84).

25 Steinar Andresen, 'Science and Politics in the International Management of Whales', *Marine Policy* vol. 13 (April 1989): 104. The three scientists in question were D. Chapman from the University of Washington, Sidney Holt from the FAO and K.R. Allen from New Zealand. John Gulland was then appointed at a later date.

26 Ibid., 104.

27 J. Rooum, 'Forhandlingene om reduksjon av fangstkvota I Antarktis. Den internasjonale hvalfangskommisjonen 1960–65' (The Negotiations of Reduction of the Catch Quota in Antarctica. The International Whaling Commission 1960–65), unpublished master's thesis, Institute for Political Science, University of Oslo, 1984, 361, found in Andresen, 'Science and Politics', 105.

28 The Committee of Three was able to accomplish a reduction in quotas and a ban of the blue whale catch, something the Scientific Committee was unable to do, since they made recommendations outside the purview of the IWC and because they offered quantifiable and specific advice in their recommendations.

29 William Aron, William Burke and Milton Freeman, 'The Whaling Issue', *Marine Policy* vol. 24 (May 2000): 180.

30 Ray Gambell, 'Whaling: Past, Present and Future', in *Whaling Controversy and the Rational Utilization of Marine Resources*. Tokyo: The Institute of Cetacean Research, 2002, 8. The NMP divided whale stocks into three categories – initial management stocks (IMS), sustained management stocks (SMS) and protected stocks (PS). Initial management stocks were whale species that have not been subjected to a hunt yet, and as such could withstand a higher-than-predicted catch quota using MSY. Sustained management stocks are those whale populations currently hunted but are plentiful enough to sustain a catch based on MSY catch rate. The final category, protected stocks, are those whale populations that are sufficiently depleted and unable to sustain a catch, thus whaling of such species is banned. See Patricia Birnie, *International Regulation of Whaling: From Conservation of Whaling to Conservation of Whales and Regulation of Whale-watching, Vol. 1*. New York: Oceana, 1985, 453, 461.

31 Peterson, 'Whalers, Cetologists, Environmentalists', 164.

32 Ibid., 166.

33 As one member of the Scientific Committee and former IWC Secretary observed of the NMP: 'The result was that the upper and lower probabilities for the calculations led to wide ranges in the possible catches which might be set, and continuing disagreements in the Commission. The whaling interests naturally advocated the highest possible catch to maintain the industry's profitability, and the more conservative-minded governments favoured a lower figure in the range provided by the scientists as a safer course to follow.' Gambell, 'Whaling: Past, Present and Future', 8.

34 Sidney Holt, 'Whale Mining, Whale Saving', *Marine Policy* vol. 9 (July 1985): 12.

35 Here, 'protectionist' and 'preservationist' are used as terms distinct from 'conservationist'. Conservationists allow for the exploitation of resources within limits supplied by scientific management, whereas protectionists and preservationists oppose the exploitation of a resource under any conditions whatsoever.

36 Peterson, 'Whalers, Cetologists, Environmentalists', 149.

37 Anthony d'Amato and Sundhir Chopra, 'Whales: Their Emerging Right to Life', *American Journal of International Law* vol. 85 (January 1991): 38.

38 Aron *et al.*, 'The Whaling Issue', 180.
39 Birnie, *International Regulation of Whaling*, 365.
40 Ibid., 422.
41 Peterson, 'Whalers, Cetologists, Environmentalists', 153.
42 Aron *et al.*, 'The Whaling Issue', 180.
43 Ibid., 180.
44 Komatsu and Misaki, *Whales and the Japanese*, 90. As a contemporary report of the decision notes: 'Japan's compliance with American pressure was based on the rational calculation that the possible cost of being driven out of the US fishing zone (the yield worth 130 billion yen and the employment of 12,000 people) was much bigger than the prospective benefit of continuing the condemned practice (11 billion yen and 1300 people).' Renpei Komatsu, 'Kujira to keizai masatsu' (Whales and Economic Friction), *Chūō Kōron* (April 1986): 89–90.
45 Interviews with Iwadō Toshiyuki, Director, International Fisheries Division, Economics Bureau, MOFA, 26 February 2004, and Nakajima Keiichi, President, Japan Whaling Association, 18 February 2004.
46 International Court of Justice, 'Whaling in the Antarctic,' (Australia v. Japan: New Zealand intervening) – Judgment of 31 March 2014. Decision can be found at www.icj-cij.org/.
47 *The New York Times*, 'U.S. to Move against Japan over Whales', 12 September 2000. For the 2005–06 season, however, Japan has more than doubled its quota to 935 minke whales and also included, for the first time, a catch of fin and humpback whales in the South Pacific. BBC News, 'Japan's Whaling Fleet Sets Sail', 8 November 2005.
48 BBC News, 'End Whaling Ban for Whales' Sake', 6 July 2001.
49 Institute of Cetacean Research, 'How Should We Effectively Conduct Research on Whale Resources?' circular (1994?).
50 See for example, 'Scientific Permits: Information on Scientific Permits, Review Procedure Guidelines and Current Permits in Effect', at IWC website, www.iwcoffice.org/conservation/permits.htm. The guidelines make it clear that lethal sampling methods will be used only when such data are unobtainable by non-lethal means.
51 Japanese Fisheries Agency, 'Current Findings of the Japanese Whale Research Program under the Special Permit in the Antarctic', *Riches of the Sea*, circular, 1995.
52 d'Amato and Chopra, 'Whales', 55.
53 Interview with Ōsumi Seiji, Director General, ICR (2 March 2004).
54 *The New York Times*, 'Yuk! No More Stomach for Whales', 24 May 2002.
55 Interview with Nakajima Keiichi, President, JWA, 18 February 2004.
56 International Whaling Commission, 'Revised Management Procedure: The Most Rigorously Tested Management Procedure for a Natural Resource Yet Developed', circular, May 2004.
57 Norway currently employs the RMP as a means to determine its own catch limits for Minke whales and seal hunting in the North Atlantic outside of IWC jurisdiction. It is being used on a trial basis to test the integrity of the management system for further use elsewhere. It also serves to challenge the political process at the IWC by illustrating how such a management system can work if the political will to enact it existed in the IWC. See David Caron, 'The International Whaling Commission and the North Atlantic Marine Mammal Commission: The Institutional Risks of Coercion in Consensual Structures', *American Journal of International Law* vol. 89 (January 1995): 165.
58 Gambell, 'Whaling: Past, Present and Future', 12.
59 IWC, 'Revised Management Scheme: Information on the Progress of RMS, International Whaling Commission', www.iwcoffice.org/conservation/rms.htm.

156   *Epistemic norm formation and whaling policy*

60 BBC News, 'Japan's Whale-Seeking Satellite', 8 January 2002.
61 BBC News, 'Whaling Safe for a Century', 4 October 2001.
62 For example, the Soviets had a quota of 720 humpback whales in the 1959/60 season, but their fleets caught a total of 12,945 – twenty times as many – approximately the total number that are thought to live in the Southern Hemisphere today. BBC News, 'Minke Whale Numbers "Declining"', 3 July 2000.
63 Institute of Cetacean Research, 'The Facts about Whales and Fish Stocks', circular, 2000.
64 BBC News, 'Japan Admits Trading Whale Votes', 18 July 2001.
65 BBC News, 'Angry Split at Whaling Meeting', 23 July 2001.
66 *Observer*, 'Save the Whales? Not if Japan's Bribes Pay Off', 13 May 2001.
67 Greenpeace, 'Rigging the System: How Japan Is Buying Control of the IWC', circular, December 2001.
68 Interview with officials at the Fisheries Agency, International Affairs Division, 23 February 2004; and the Japan Whaling Association, 18 February 2004.
69 See for example, d'Amato and Chopra, 'Whales', passim.
70 See for example Kalland and Moeran, *Japanese Whaling*, and *The 1st Summit of Japanese Traditional Whaling Communities*, passim.
71 Ministry of Foreign Affairs, 'Japan and the Management of Whales', www.mofa. go.jp/policy/.
72 Interview with Nakajima Keiichi, President, Japan Whaling Association, 18 February 2004. It is also important to remember that the taste for whale meat acquired by many middle-aged Japanese was largely the product of the SCAP Occupation authorities' promotion of serving whale meat in school lunches in the early post-war period. Indeed, many parts of Japan did not have a tradition of eating whale meat at all prior to the Occupation and thus the promotion of whale meat in the Japanese diet can be seen as a further example of the imposition of foreign norms upon the Japanese. Scheiber, *Inter-Allied Conflicts*, 136. Also refer to Milton M.R. Freeman, 'The Historical Legacy of Industrial Whaling and Current Problems in Japan's Coastal Fishery', in *Ocean Resources: Industries and Rivalries Since 1800*, Harry Scheiber, ed. Berkeley, CA: Center for the Study of Law and Society, University of California, 1990.
73 Kalland and Moeran, *Japanese Whaling*, 182–92.
74 Aron *et al.*, 'The Whaling Issue', 181.
75 Many of the officials interviewed spoke of the Japanese appetite for whale meat as a feature of Japanese culture that makes Japan distinct from other nations. This may explain why the moratorium engenders a heated response from Japanese nationalists who see the banning of whale meat consumption on the same level as cultural annihilation.
76 The whaling issue is also adopted by some advocates of *Nihonjinron* (theory of Japanese uniqueness), who see Japan's tradition of eating whale meat as distinguishing the Japanese as a nation or race from their European counterparts (although ignoring that the practice exists, for example, in both Norway and Iceland). See Tetsuo Hiraguchi, 'Prehistoric and Protohistoric Whaling, and Diversity in Japanese Foods', in *The 1st Summit of Japanese Traditional Whaling Communities*. Tokyo: Institute of Cetacean Research, 2002, 23–47, for an interesting example of the use of nihonjinron in archaeology.
77 Shūgiin norinsuisan iinkai kaigiroku (Minutes of Agriculture, Forestry and Fisheries Committee, House of Representatives), 28 July 1987. Found in Isao Miyaoka, 'International Norms and State Autonomy: Wildlife Preservationist Pressures on Japanese Economic Practices, 1987–91', DPhil thesis, University of Oxford, 1998, 148–49.
78 BBC News, 'Japan Campaigns for Whaling', 19 June 2000.
79 BBC News, 'Japanese People Support Whaling', 16 March 2002.

80 *The New York Times*, 'Yuk! No More Stomach for Whales', 24 May 2002.
81 Masayuki Komatsu, a senior official in the Fisheries Agency, regularly makes such public comments. See for example, BBC News, 'Japan Threatens Whaling Walkout', 16 June 2003. Similar discussions also came up during talks with officials at the Japan Fisheries Association, 12 February 2004.
82 Interview with Iwadō Toshiyuki, Director, International Fisheries Division, Economics Bureau, MOFA, 9 March 2005.
83 *The Japan Times*, 'Maintaining IWC Membership is in Japan's Interest', 9 November 1994.
84 Here, 'protectionist' and 'preservationist' are used as terms distinct from 'conservationist'. Conservationists allow for the exploitation of resources within limits supplied by scientific management, whereas protectionists and preservationists oppose the exploitation of a resource under any conditions whatsoever.
85 BBC News, 'Japan Pushes for Whale Meat Revival', 19 June 2005.
86 The February 2004 conference on whaling sponsored by the Ministry of Foreign Affairs was entitled 'Towards the Mutual Understanding between Sustainable Use and Protection of Marine Resources'. The conference came out in favour of re-opening whaling using 'sustainable-use' as the core management concept.
87 Quote found in Peter Stoett, *The International Politics of Whaling*. Vancouver: University of British Columbia Press, 1997, 77.
88 Institute for Cetacean Research, 'Introduction', www.icrwhale.org/abouticr.htm.
89 Kazuo Shima, 'Whaling and the Rational Use of Living Resources', *Whaling Controversy and the Rational Utilization of Marine Resources*. Tokyo: The Institute of Cetacean Research, 2002, 28.
90 Ethan Nadelmann, 'Global Prohibition Regimes: The Evolution of Norms in International Society', *International Organization* vol. 44 (Autumn 1990): 523.
91 Michael Berril, *The Plundered Seas: Can the World's Fish Be Saved?* Vancouver: Greystone Books, 1997, 42.
92 Such scepticism is particularly severe in the case of whaling, perhaps owing to the dominant position that business occupied over science during the early years of the IWC. Fisheries managers held sway in the commission and scientists were unable to bring to bear the same political pressure that managers possessed to influence policy outcomes. In this epistemic vacuum, it was the economics of industry's bottom line rather than ecosystem science that determined catch levels of whales. When the environmental movement achieved a degree of power and prominence in the 1970s, arguments among whalers, cetologists and environmentalists tended to centre around the interpretation of 'uncertainty' in science. The environmentalists and more cautious cetologists interpreted uncertainty as a mandate for severe limits of whaling and argued that, because of this uncertainty, all stocks should be treated as equally endangered and left to recover.
93 See, for example, Greenpeace International's 'Save Our Seas' campaign that targets over-fishing as a new environmental threat, www.greenpeace.org/international_en/campaigns/.
94 BBC News, 'Bitterness Clouds Summit Finale', 4 September 2002.
95 See World Wildlife Fund, worldwildlife.org/oceans/threats.cfm; or Greenpeace, www.greenpeace.org/international_en/campaigns/.
96 Food and Agriculture Organization, *White Paper on World Fisheries*. Rome: Food and Agriculture Organization, 2000.
97 World Conservation Union, www.redlist.org.
98 Reg Watson and Daniel Pauly, 'Systematic Distortions in World Fisheries Catch Trends', *Nature* vol. 414 (29 November 2001): 536.
99 The authors suggest such falsification occurred since Chinese managers are promoted on the basis of reported fish catches and thus had an incentive to over-

report figures in the interest of their careers. Watson and Pauly, 'Systematic Distortions', 535.
100 Ransom A. Myers and Boris Worm, 'Rapid Worldwide Depletion of Predatory Fish Communities', *Nature* vol. 423 (15 May 2003): 280.
101 Daniel Pauly, Villy Christensen, Johanne Dalsgaard, Rainer Froese and Francisco Torres, Jr, 'Fishing Down Marine Food Webs', *Science* vol. 279 (6 February 1998): 860.
102 Charles Clover, *The End of the Line: How Over-Fishing is Changing the World and What We Eat*. London: Ebury Press, 2004.
103 Clover, *The End of the Line*, 3.
104 Interviews with Director, International Fisheries Division, Economics Bureau, MOFA, 9 March 2005; and Director, International Affairs Division, Fisheries Agency, 23 February 2004.
105 Komatsu and Misaki, *Whales and the Japanese*, 165–66.
106 Interview with Director, International Fisheries Division, Economics Bureau, MOFA, 26 February 2004.
107 Interview with Director, International Affairs Division, Fisheries Agency, 23 February 2004.
108 Interview with President, Japan Whaling Association (JWA), 18 February 2004.
109 Interview with President of JWA. An experimental lunch programme that serves whale meat has already been tested in Wakayama Prefecture in 2005. BBC News, 'Japan Pushes for Whale Meat Revival', 19 June 2005.

Bibliography

Andresen, Steinar. 'Science and Politics in the International Management of Whales', *Marine Policy* vol. 13 (April 1989): 99–117.

Aron, William, William Burke and Milton Freeman. 'The Whaling Issue', *Marine Policy* vol. 24 (May 2000): 179–91.

Berril, Michael. *The Plundered Seas: Can the World's Fish Be Saved?* Vancouver: Greystone Books, 1997.

Birnie, Patricia. *International Regulation of Whaling: From Conservation of Whaling to Conservation of Whales and Regulation of Whale-watching, Vol. 1 and 2*. New York: Oceana, 1985.

Clover, Charles. *The End of the Line: How Over-Fishing is Changing the World and What We Eat*. London: Ebury Press, 2004.

d'Amato, Anthony and Sundhir Chopra. 'Whales: Their Emerging Right to Life', *American Journal of International Law* vol. 85 (January 1991): 21–62.

Freeman, Milton. 'The Historical Legacy of Industrial Whaling and Current Problems in Japan's Coastal Fishery', in *Ocean Resources: Industries and Rivalries Since 1800*, Harry Scheiber, ed. Berkeley, CA: Center for the Study of Law and Society, University of California Berkeley.

Gambell, Ray. 'Whaling: Past, Present and Future', in *Whaling Controversy and the Rational Utilization of Marine Resources*. Tokyo: The Institute of Cetacean Research, 2002.

Hirasawa, Yutaka. 'The Whaling Industry in Japan's Economy', in *The Whaling Issue in U.S.-Japan Relations*, John R. Schmidhauser and George Totten III, eds. Boulder, CO: Westview Press, 1978.

Kalland, Arne and Brian Moeran. *Japanese Whaling: End of an Era?* London: Curzon Press, 1992.

Komatsu, Masayuki and Shigeko Misaki. *Whales and the Japanese*. Tokyo: Japan Whaling Association, 2003.

Myers, Ransom A. and Boris Worm. 'Rapid Worldwide Depletion of Predatory Fish Communities', *Nature* vol. 423 (15 May 2003): 280–83.

Nadelmann, Ethan. 'Global Prohibition Regimes: The Evolution of Norms in International Society', *International Organization* vol. 44 (Autumn 1990): 479–526.

Pauly, Daniel, Villy Christensen, Johanne Dalsgaard, Rainer Froese and Francisco Torres, Jr. 'Fishing Down Marine Food Webs', *Science* vol. 279 (6 February 1998): 860–63.

Peterson, M.J. 'Whalers, Cetologists, Environmentalists, and the International Management of Whaling', *International Organization* vol. 46 (Winter 1992): 147–86.

Scarff, J.E. 'The International Management of Whales, Dolphins and Porpoises: An Interdisciplinary Assessment', *Ecology Law Quarterly* vol. 6 (1977): 323–427.

Stoett, Peter. *The International Politics of Whaling*. Vancouver: University of British Columbia Press, 1997.

Takahashi, Junichi. 'Whaling Culture in Contemporary Japan', in *The 1st Summit of Japanese Traditional Whaling Communities*. Tokyo: Institute of Cetacean Research, 2002.

Tonnessen, J.N. and A.D. Johnsen, *The History of Modern Whaling*. Berkeley, CA: University of California Press, 1982.

Watson, Reg and Daniel Pauly. 'Systematic Distortions in World Fisheries Catch Trends', *Nature* vol. 414 (29 November 2001): 534–36.

9 Food security and self-sufficiency today

Given the goals agreed upon in Japan's comprehensive security strategy and the increasingly restrictive nature of the international ocean regime since the mid-1970s, what are the current trends in domestic food policy in Japan today? What are the presently stated goals of Japanese food policy and upon what foundations have they been developed? How does Japan compare with other developed nations with regard to self-sufficiency?

The Food Control Law of 1942 was finally replaced in July 1999 by the Basic Law on Food, Agriculture and Rural Areas. The Basic Law aims to ensure a stable food supply at reasonable prices: 'in consideration of the fact that there are certain unstable factors in the world food trade and supply/demand, this stable food supply to the people shall be secured with increase of domestic agricultural production as a basis, together with an appropriate combination with imports and stockpiles.'[1] It also states that 'even in the case that domestic supply is insufficient to meet demand or is likely to be for a certain period, due to unexpected situations such as a bad harvest or interrupted imports, the minimum food supply required for the people shall be secured in order not to be a hindrance to the stability of people's lives and smooth operation of the national economy'.[2] The Basic Law also establishes a plan that, among other things, sets a target for the food self-sufficiency ratio 'in view of improving the ratio and as a guideline for domestic agricultural production and food consumption'.[3] Self-sufficiency has remained a keystone for the administration of Japanese food policy in spite of the opportunity presented by the introduction of new legislation to amend the food control system after more than fifty years of operation.

Moreover, public opinion resoundingly supports the government's position on self-sufficiency. In October 2000, the Public Opinion Research Section within the Prime Minister's Office released the results of a study entitled 'Public Opinion Survey in Regard to Trade in Agricultural Products' (*Nōsanbutsu bōeki nikan suru yoron-chōsa*).[4] When asked whether they would choose domestic or imported products that were otherwise identical, 82 percent of respondents answered that they would prefer domestic products, whereas only 17 percent said they did not care whether the products were foreign or Japanese.[5]

The survey also found that 44 percent of respondents thought it was better to produce foodstuffs in Japan as much as possible while lowering domestic production costs, even if foodstuffs are more expensive than foreign products. An additional 41 percent felt that 'even if they are more expensive than foreign products, at least for main staples such as rice it is better to produce as much as possible in Japan while lowering domestic production costs'.[6] Only 11 percent responded that it is better to import those foodstuffs where the foreign product is cheaper. Similarly, when asked about what policy Japan should emphasize at World Trade Organization talks regarding agricultural subsidies, 72 percent of respondents said that 'ensuring a guaranteed stable supply of foodstuffs' was most important.[7] The Ministry of Agriculture, Forestry and Fisheries (MAFF) has used the results of this survey as public acceptance of higher agricultural product prices to protect domestic production and stability of supply under the policy of self-sufficiency.

Thus, it comes as no surprise that considerable national policies and mechanisms are in place to maximize self-sufficiency to the greatest extent possible. A significant change in eating habits over the decades has led to a significant decline in self-sufficiency rates, which can help explain the reasons why Japanese policy makers have been concerned with food security and alarmed at recent trends.

Figure 9.1 shows that no matter which formula is used for calculating self-sufficiency, the rate has been in steady decline throughout the post-war period. The most common measurement used by analysts within the Japanese government and academia is the calorie ratio, which, according to Figure 9.1, fell from a high of 73 percent self-sufficiency in 1965 to 39 percent in 2011.[8] Japan's rate of self-sufficiency is currently calculated among the lowest among

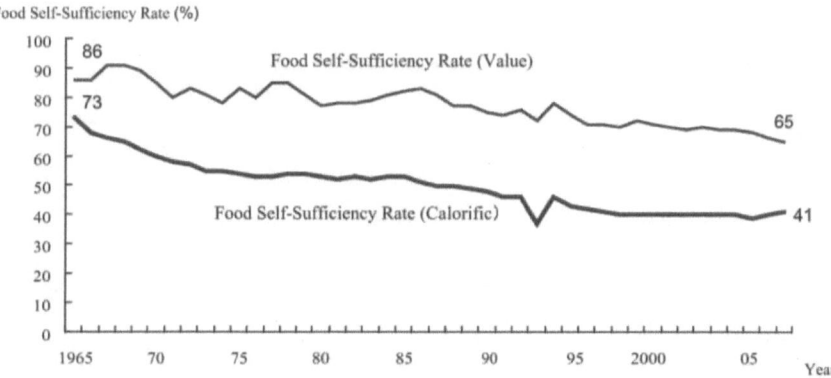

Figure 9.1 Trends in Japan's food self-sufficiency ratio
Source: Ministry of Agriculture, Forestry and Fisheries, *Heisei 23-nendo shokuryo jik-kyū-ritsu wo meguru jijō* (Situation Concerning the Self-Sufficiency Rate in 2011). Tokyo: MAFF, 2011

developed nations, compared to 173 percent for Australia, 168 percent for Canada, 111 percent for France and 124 percent for the United States.[9]

In the case of fisheries, the self-sufficiency rate stands at 50 percent, which is higher than the national agricultural average.[10] Unlike agriculture, however, there are few variables in domestic production that will enable this figure to increase in the future. In other words, the most readily available and productive fishing grounds (found in Japanese exclusive economic zone, or EEZs) are already fished to maximum effort and further increases are very unlikely to be seen. Thus any sizeable increase in appetite for fish will have to be met with increasing imports.

In order to improve upon the self-sufficiency rate for food in general, MAFF has targeted two new areas for improvement. These are to increase the amount of land available for agricultural purposes through re-zoning, reclamation or conversion, as well as an effort to have more youth enter agriculture as a profession. Currently, agricultural land only accounts for 0.04 percent of all land usage in Japan, which is extremely low even compared to a geographically similar country such as New Zealand (4.41 percent) or a very mountainous country, such as Switzerland (0.21 percent).[11] Under the Basic Plan 2005, new agricultural policy targets government assistance to farmers who satisfy certain conditions, especially minimum farm size, to compel farmers to expand the size of farm operations.[12] Moreover, the government aims to increase the number of families involved in agricultural production, since the workforce is ageing and decreasing, with a sizeable portion of the agricultural population in the elderly 60–75 age range (see Figure 9.2).

According to MAFF, a change in eating habits accounts for nearly 70 percent of the self-sufficiency rate's decline.[13] Consumption of foods such as meat for which the self-sufficiency rate is low has increased considerably in

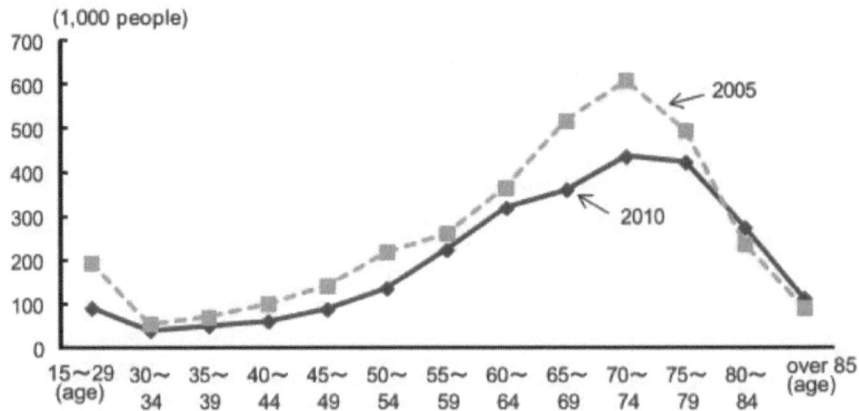

Figure 9.2 Changes in population engaged in farming by age (nationwide)
Source: Official Statistics of Japan, *Digest on the Results of 2010 World Census of Agriculture and Forestry*, www.e-stat.go.jp

recent years. For example, the annual per capita consumption of rice dropped by 40 percent from 112 kg in fiscal year 1965 to 67 kg in fiscal year 1997, while the per capita consumption of meat increased by more than three times, from 9 kg to 31 kg over the same period, and that of oils and fats jumped from 6 kg to 15 kg. As a result, feed grain imports rose by nearly three times, to 16 million metric tons, and imports of the raw materials for oil, such as soybeans, also increased.[14] Thus, while the consumption of rice has declined over the years despite its high self-sufficiency rate, the consumption of live-stock products, which depend on imported feed and raw materials, has been increasing.

It should be noted that despite a relatively low level of self-sufficiency, Japan has met its food needs every year since returning to self-governing status in 1952. Even though the government has designed contingency plans to deal with food shortages in the event of natural disaster, international hostilities or worldwide food shortages, none of these scenarios has come to pass in the post-war period to any level that might threaten Japan's food supply. In fact, as an advanced nation with vast financial resources, Japan seems likely to be able always to purchase the food resources it needs from abroad even during times of emergency – albeit at higher costs than before – and is most certainly better off than developing countries that do not carry such financial clout.

A rather important distinction should be made with respect to a govern-ment's stated aims and the actual outcomes. The scale of activities relating to food control and distribution as well as the independence of vital variables can place a large gap between planning and results. In short, a government may set admirable targets for national growth and consumption, but actual consumption patterns or the facts of geography can still be overriding factors in national outcomes.

For example, owing to record low levels of self-sufficiency, MAFF set new targets in 2000 for improving self-reliance in the coming century. It used the Food Balance Sheet to assess general food supply and demand in Japan in order to identify the total volume of food nutrition supplied in Japan from their production to final consumption, and to indicate the per capita net food supply and nutritional volume.[15] Based upon these assessments, a new target was set to increase to a level of 45 percent self-sufficiency by the year 2010, which is an increase of 5 percent over ten years.[16]

MAFF made a number of recommendations to this effect, but no concrete plan was imposed and 'guidance' rather than 'edict' was the effective form of implementation. Many people, both within the ministry and outside, remained sceptical that such an increase could have been achieved by 2010, but yet officials quoted these targets when explaining the rationale behind many fisheries policies.[17] The actual self-sufficiency rate achieved in 2011 was 39 percent (calorific, or 66 percent according to price), which was a sizeable shortfall from the stated aims of the ministry.[18] Naturally, the government cannot control for all the variables that go into the production and distribution of food, while its 'control' of the actual appetites of consumers nationwide is even less so. In fact, the outcomes of food security policy have followed the pattern outlined in the Inoki Task Force Report from 1978 in that it appears that the national food balance offers a healthy mix of domestic production and imports that help diversify the market.

Challenges posed by natural and manmade disasters: Fukushima crisis

On 11 April 2011, a 9.1 magnitude earthquake struck the Tohoku region, which then triggered a devastating tsunami and contributed to the meltdown of a number of reactors in the Fukushima area, including the Dai-ichi nuclear reactors located on the coast of Futaba City, Fukushima. The explosion and subsequent meltdown of reactors emitted a sizeable amount of radiation to the extent that it was classified a Level 7 Event on the International Atomic Energy Agency scale – the highest possible hazardous rating, on par with the Chernobyl disaster of 1984 – and led to an evacuation of many towns in Fukushima Prefecture. Some 370,000 terabecquerels of radioactive iodine and caesium were released, which was considerably more than Tokyo Electric Power Corporation (TEPCO) officials had originally thought.[19]

Initially, the radiation cloud spread far and wide, contaminating a largely rural area with alarming levels of radioactive caesium and iodine that found its way into the water and food chain. Moreover, seawater was used as an emergency coolant for reactors that had gone out of control during the early stages of the crisis and the waste water was dumped directly into the Pacific Ocean where the Kuroshio current runs northward into Japan's largest fishing grounds located off the coast of Miyagi Prefecture.

Apart from the immediate evacuation measures that were taken to deal with the crisis, additional measures were adopted to prevent the consumption of potentially radioactive foodstuffs. In a country where food safety is a priority, the designation of banned agricultural products and areas has been quite controversial, but officials have implemented various testing measures and standards to ensure scientifically acceptable levels of radiation for the safe consumption of food. Japanese authorities have compiled data on the concentrations of radioactive caesium on the ground surface at roughly 2,200 locations within approximately 100 km from the Fukushima Dai-ichi nuclear plant, using air dose rates and collecting soil samples (see Figure 9.3).[20]

Certain radiation isotopes have a long life (caesium Cs-137 has a thirty-year half-life) and will remain present in the ecosystem for some time to come, which has raised concerns about food safety for agricultural and fisheries products from the Tohoku area. By May 2011, the government had tested a range of agricultural products for safety levels and had determined that over 93 percent of their results proved to have safe levels of radioactive iodine and caesium.[21] Milk, leafy vegetables and other smaller food sources were given temporary restrictions, mostly affecting the area of Fukushima

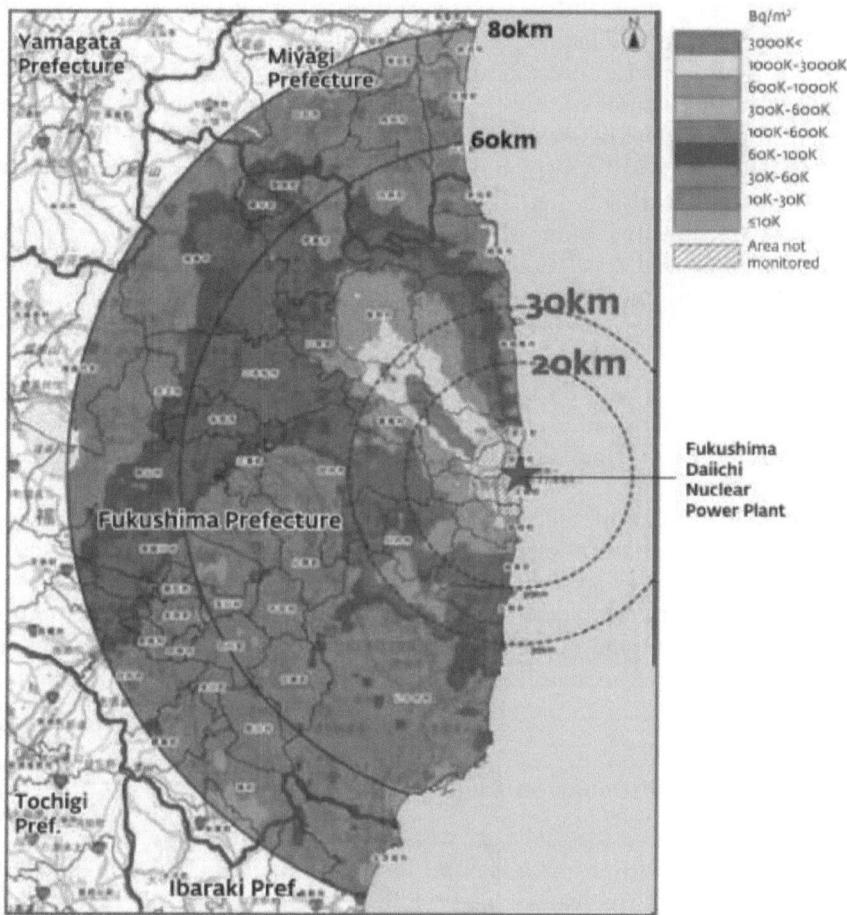

Figure 9.3 Map showing accumulated radioactive caesium levels, July 2011
Source: National Diet of Japan, *The Official Report of the Fukushima Nuclear Accident Independent Investigation Commission, Executive Summary.* Tokyo: National Diet of Japan, 2011, 40

Prefecture, but these were removed within the first half-year after the incident, although monitoring still exists for the Tohoku area. Perhaps the greatest impact on agricultural production in the Fukushima and Tohoku areas in the future will be the stigma attached with radiation poisoning and distrust of food products from that area, which may take years for consumers to overcome. Many areas surrounding the Dai-ichi nuclear reactors remain evacuation zones that have since become ghost towns, the images of which still haunt the media.

Fisheries have also been impacted by the nuclear crisis. Alarming levels of radioactive seawater were pumped into the ocean during the emergency phase of the Dai-ichi plant meltdown, which spiked radiation levels in the area.[22] It is expected, however, that the great quantity of water in the Pacific Ocean would rapidly disperse and dilute radioactive materials. Since contaminated water tends to accumulate at the bottom of the ocean, and the Kuroshio current runs northward toward fish spawning grounds off the Miyagi coastline, some concern still exists, however, with regard to the long-term impact on fisheries in the area.

Much as with the case of agricultural products, Japanese regulatory authorities imposed regulation limits for radionuclides on fisheries products and subsequently monitored fish caught in the prefectures surrounding the damaged nuclear power plants. It was found that only Japanese sand-lance exceeded such levels and were subject to a ban, but all other species passed testing, including migratory species that traverse the affected waters.[23] Again, it would appear that the stigma attached to radiation – and not any kind of regulatory ban – will have the greatest impact on consumption patterns for fisheries in the area.

Although the size of this natural calamity and subsequent radiation disaster had the potential to impact Japanese food security, it would appear that any potential long-term impact was minimized through close radiation monitoring and, when the need arose, through the provision of alternative sources of supply (mostly domestic). Of course, the impact of radiation exposure and poisoning take time to manifest, so this will remain an issue of concern in years, if not generations, to come.

Impact of self-sufficiency on trade relations

Naturally, when food security policy is devised in a nation where the political clout of the agricultural sector is enormous and its economy is fundamentally based upon trading networks, there will be politicization of relations with trading partners where agricultural products are concerned. Some critics have accused Japan of adopting a 'beggar thy neighbour' agricultural policy in relation to poorer Asian nations, since Japan's attempts to foster self-sufficiency create barriers to trade and market development in agricultural products with which developing nations have a comparative advantage and Japan is distinctly uncompetitive.[24] Simple Ricardian logic may suggest that as a nation well endowed with technological and scientific skills, Japan's comparative advantage may best lie in applying its workforce in higher-scale industries and not through the promotion of its own agricultural sector.

These arguments are likely to be reappraised in the forthcoming talks on the Trans-Pacific Partnership (TPP), now that Japan has been included as a full member nation.[25] Agricultural policies will likely feature in upcoming TPP talks which will likely help determine the Asia-Pacific agenda in the coming years. Many political parties came out in opposition to including agriculture on the TPP agenda, but the government has agreed to join the TPP negotiations in any event and has established a ministerial committee to evaluate negotiation policy for the upcoming round of talks.[26] Surely the question of comprehensive security will yet again reach national proportions and the TPP talks may offer yet another chance for a re-evaluation of agricultural trade policy.

It should also be noted that Japan is not unique with respect to having a fully developed economy that heavily subsidizes its own agricultural sector. In fact, most advanced industrial nations have high levels of government support for their rural communities, and these transfers are frequently couched in terminology relating to national sustenance, health and food security.[27] What makes Japan unique in this respect is that its food security policy is intimately linked with its *defence* policy.

One nation well known for farming subsidies is France. Nominal rates of government assistance to all products, both meant for export and domestic consumption, was 11 percent in 2007, 33 percent in 2000, 66 percent in 1990 and as high as 118 percent in 1987. While this suggests a declining overall trend for agricultural subsidies in recent years, the rate is still comparatively high considering that France is a high-income country serving at the forefront of the European Economic Community.[28] In a term coined by Ben Clift of Warwick University, UK, this proclivity to offer rural subsidies in high-income countries may be best described as 'economic patriotism'. This is a natural outgrowth of a situation when a state must balance its commitments to the free market with the encouragement of business on its home territory while sustaining social order.[29] The Common Agricultural Policy is less security focused than Japanese food security, but still equally infused with patriotic sentiment, as a means to support rural development and income assistance to farmers who must compete with lower prices on the world market.

Comprehensive security today

Given that TPP talks will likely result in a reconsideration of comprehensive security fundamentals, especially considering that negotiations will involve questions of trade and agriculture for the Japanese negotiation team, a few questions are now quite pertinent for national policy. How solid are the foundations of comprehensive security today and does it serve any longer as a rationale for policy making in domestic and foreign realms? Is it a realistic goal any longer given that its operational scope and capacity is premised on the US-Japan joint security arrangement that is currently under re-evaluation?

Comprehensive security is still a fundamental operating principle for many ministries in the Japanese government and is widely understood by the citizenry and politicians to be a concept linked to Japan's very survival in a turbulent and sometimes hostile global environment. In the Ministry of Foreign Affairs (MOFA) *Blue Book*, 'comprehensive' approaches to security issues and human security still feature prominently in the context of gathering powers and capacities from multiple sources in order to problem solve.[30] In a recent agreement with India, MOFA deliberately chose the term 'Comprehensive Security Dialogue' to frame its consultations, especially since India was reconsidering the basis of its own security arrangements given new geo-strategic realities that predominate in South Asia and its own ambitions vis-à-vis permanent membership of the United Nations (UN) Security Council.[31]

Japan has been an active proponent of world food security abroad, especially as it pertains to the UN mandate and the Millennium Development Goals. Frequently, the policy is presented within the context of human security, such that fundamental human rights, individual-level security and the conditions required for development are difficult to achieve without the premium afforded by having food security for individuals, homes and communities.

This diplomatic push had its greatest presence in the early 2000s as various UN organizations, including the Food and Agriculture Organization (FAO), reconstrued policy aims in favour of individual, not strictly state, levels of security provision. The mutual conceptualization linking both human and food security was frequently termed in diplomatic agreements as an understanding on the notions of 'comprehensive security'. It might also be noted that the promotion of food security worldwide also coincided with the greater push abroad by the People's Republic of China, Europe and North America for better access to resource agreements in the mid-2000s.

Conclusion

This chapter started out by asking a number of questions relating to Japan's comprehensive security policy and food security today: What are the current trends in domestic food policy and challenges they might face in the future? What are the presently stated goals of Japanese food policy upon which national goals, especially pertaining to rural development, have been based? How does Japan compare with other developed nations with regard to self-sufficiency?

Japan is in a situation where its self-sufficiency rate is in steady decline, but at a rate that is sustainable and understandable given its geo-economic predicament. Food security is likely to be resilient since Japanese farms are quite productive, alternative supplies exist and even a tragedy on the scale of the Fukushima disaster did not destabilize the food control system. Moreover, official commitment to comprehensive security has also entailed several policy initiatives internationally that are commendable, such as the adoption of food security and human security agendas on the global stage, in particular with UN agencies. This may be seen as a direct contribution made by Japanese diplomats internationally. Lastly, comprehensive security remains an all-encompassing national policy that is both congruent with the Yoshida Doctrine and a distinctly Japanese approach to address the insecurities faced as a nation not endowed with abundant natural resources.

Notes

1 Ministry of Agriculture, Forestry and Fisheries, *Basic Law on Food, Agriculture and Rural Areas*, Article 2, www.maff.go.jp/soshiki/kambou/kikaku/NewBLaw/BasicLaw.html.
2 Ibid., Article 2.
3 Ibid., Article 15.
4 Fieldwork for the study was conducted in July 2000 and involved personal interviews with a sample of 5,000 randomly selected Japanese nationals aged twenty years and above. A 71 percent completion rate was achieved, yielding 3,570 completed interviews (1,644 males and 1,926 females) (J@pan Inc. Magazine, www.japaninc.net/mag/comp/2001/03/mar01_reports_js.html).
5 Ibid.
6 The caveat to the question – while lowering domestic production costs – is problematic and some analysts have suggested that it is misleading to the public, skewing the results in favour of domestic production. The results would likely be quite different if respondents were informed of the costs involved in supporting domestic production of foodstuffs. Ibid.
7 Ibid.
8 The calorie ratio is determined by converting all foods and livestock feed into calories, which are then calculated into domestic demand minus exports that are supplied by a ratio of domestic production and imports.
9 Ministry of Agriculture, Forestry and Fisheries, *Information Sheet on the Self-Sufficiency Rate, 2011*, www.maff.go.jp/chushi/jikyu/pdf/240207jikyupanfu.pdf.
10 Ministry of Agriculture, Forestry and Fisheries, *Heisei 23-nendo shokuryō jikkyū-ritsu wo meguru jijō* (Situation Concerning the Self-Sufficiency Rate in 2011). Tokyo: MAFF, 2011, appendix.
11 Food and Agriculture Organization (FAO), *The State of Food and Agriculture: Food Aid for Food Security*. Rome: FAO, 2006, 137.
12 Masayoshi Honma and Yujiro Hayami, 'Distortions to Agricultural Incentives in Japan, Korea and Taiwan', *Agricultural Distortions Research Project Working Paper XX*. New York: World Bank's Development Research Group, October 2006, 10.
13 Ministry of Foreign Affairs, 'Trends in Japan', *Japan Echo*, web-japan.org/trends00/honbun/tj000604.html.
14 Ibid.
15 The Food Policy Planning Division, General Food Policy Bureau, Ministry of Agriculture, Forestry and Fisheries publishes the Food Balance Sheet (*Shokuryō jikyū-hyō*) every fiscal year based on FAO statistics in order to monitor and evaluate self-sufficiency rates on a yearly basis. The 2000 publication is published here: Ministry of Internal Affairs and Statistics, Statistics Bureau, www.stat.go.jp/english/index/official/205.htm#11.
16 Ministry of Agriculture, Forestry and Fisheries, *Shokuryō-nōgyō-nōson kihon kei-kaku, Heisei 12-nen 3-gatsu* (Basic Plan for Food Supply, Agriculture and Agricultural Villages, March 2000). Tokyo: Ministry of Agriculture, Forestry and Fisheries, 2000.
17 Interviews with officials from MOFA and MAFF, February 2004/March 2005.
18 Ministry of Agriculture, Forestry and Fisheries, *Heisei 23-nendo shokuryō jikkyū-ritsu wo meguru jijō* (Situation Concerning the Self-Sufficiency Rate in 2011).
19 Eliza Barclay, 'Fukushima vs. Chernobyl: Still Not Equal', *National Public Radio (NPR)*, 12 April 2011.
20 International Atomic Energy Agency, *Fukushima Daiichi Status Report, October 27, 2011*. Vienna: IAEA, 2011, 2.

21 International Atomic Energy Agency, Fukushima Nuclear Accident Update Log, 2 June 2011, www.iaea.org/newscenter/news/tsunamiupdate01.html.
22 World Health Organization (WHO), 'Impact on Seafood Safety of the Nuclear Accident in Japan', *Information Sheet*, 11 May 2011. New York: WHO, 2011, 1. Testing of marine water 30km off the coast of Japan has shown that the concentrations of radionuclides have dropped rapidly to very low levels.
23 Ibid., 2.
24 Yujiro Hayami, 'Food Security: Fallacy or Reality?' *Food Security in Asia: Economics and Policies*, Wen Chern, Colin Carter and Shun-Yi Shei, eds. Cheltenham, UK: Edward Elgar, 2000, 16–17.
25 Government of Japan, Prime Minister's Office, Press Briefing, 5 April 2013.
26 Government of Japan, Prime Minister's Office, News Release, 12 April 2013, www.kantei.go.jp/foreign/96_abe/actions/201304/12tpp_e.html.
27 In terms of nominal rates of assistance (NRA) to measure agricultural subsidies and price effects in 2007, Japan (104) is much higher relative to other advanced economies, such as France (13), United States (7) and New Zealand (1). This has more to do with price effects rather than absolute terms of subsidy provision. Ernesto Valenzuela, Johanna Croser, Esteban Jara, Signe Nelgen and Kym Anderson, 'Annual Estimates of Distortions to Agricultural Incentives in High-Income Countries', *Agricultural Distortions Working Paper 74*. New York: World Bank Development Research Group, September 2008, *intra*.
28 Valenzuela *et al.*, 'Annual Estimates of Distortions', 22.
29 Ben Clift and Cornelia Woll, 'The Revival of Economic Patriotism', in Glenn Morgan and Richard Whitley, eds. *Capitalism and Capitalisms in the 21st Century*. Oxford: Oxford University Press, 2011, 70.
30 See, for example, the Diplomatic Bluebook for 2011, wherein support for the United Nations, political and security matters, human security and food aid abroad are all to be addressed 'comprehensively'. MOFA, *Diplomatic Bluebook 2011*. Tokyo: Ministry of Foreign Affairs, 2012.
31 Ministry of Foreign Affairs, '8th Japan-India Security Dialogue', Press Release, 13 May 2011.

Bibliography

Clift, Ben and Cornelia Woll. 'The Revival of Economic Patriotism', in *Capitalism and Capitalisms in the 21st Century*, Glenn Morgan and Richard Whitley, eds. Oxford: Oxford University Press, 2011.
FAO (Food and Agriculture Organization). *The State of Food and Agriculture: Food Aid for Food Security*. Rome: FAO, 2006.
Hayami, Yujiro. 'Food Security: Fallacy or Reality?' in *Food Security in Asia: Economics and Policies*, Wen Chern, Colin Carter and Shun-Yi Shei, eds. Cheltenham, UK: Edward Elgar, 2000.
Honma, Masayoshi and Yujiro Hayami. 'Distortions to Agricultural Incentives in Japan, Korea and Taiwan', in *Agricultural Distortions Research Project Working Paper XX*. New York: World Bank's Development Research Group, October 2006.
International Atomic Energy Agency. *Fukushima Daiichi Status Report, October 27, 2011*. Vienna: IAEA, 2011.
Ministry of Agriculture, Forestry and Fisheries. *Basic Law on Food, Agriculture and Rural Areas*. n.d., www.maff.go.jp/soshiki/kambou/kikaku/NewBLaw/BasicLaw.html.
——*Heisei 23-nendo shokuryō jikkyū-ritsu wo meguru jijō* (Situation Concerning the Self-Sufficiency Rate in 2011). Tokyo: MAFF, 2011.

——*Information Sheet on the Self-Sufficiency Rate, 2011.* 2011, www.maff.go.jp/chushi/jikyu/pdf/240207jikyupanfu.pdf.

——*Shokuryō-nōgyō-nōson kihon keikaku, Heisei 12-nen 3-gatsu* (Basic Plan for Food Supply, Agriculture and Agricultural Villages, March 2000). Tokyo: Ministry of Agriculture, Forestry and Fisheries, 2000.

Ministry of Foreign Affairs. *Diplomatic Bluebook for 2011.* Tokyo: Ministry of Foreign Affairs, 2012.

WHO (World Health Organization). 'Impact on Seafood Safety of the Nuclear Accident in Japan', *Information Sheet* (11 May 2011).

Conclusion

This book focused on ocean diplomacy and politics as it pertains to Japanese maritime and fisheries policy in the Pacific during the post-war period. Through an examination of the motives and means for international fisheries policy and an analysis of the domestic and international factors that have facilitated or constrained them, a much more realistic understanding of the politics of ocean resource use and management can be achieved.

At the outset, this book stated that ocean governance may be understood as the product of three stages of interaction: 1) international legal norms and systems; 2) national interests and politics; and 3) issues and cases that raise technical questions and conceptual clashes. Although the book devotes a sizeable amount of attention to the specific case of Japanese ocean policy in the context of its comprehensive security strategy, it also highlights the importance of international law in forging the context within which international relations are conducted.

International legal context

In particular, several innovations in global legal norms and regimes arose in the latter part of the twentieth century in light of experience, conflict, negotiation and scope. The most worthy developments of note include, but are not limited to, the precautionary principle, enclosure and stewardship, the Code of Conduct for Fisheries, food security and exclusive economic zones (EEZs). Since these concepts and systems have been sufficiently elucidated in earlier chapters, the following section will offer some observations with respect to possible conflicts and innovations in international maritime law in the coming decades.

Considerable specification and complexity has advanced the cause of international law as both a guiding force for regulation of ocean use and as an arbiter in international conflicts. In the current environment, three components of international law stand out among the vast number of laws and treaties for further consideration, given their relevance for contemporary problems and usefulness in interstate negotiations. These are: 1) the United Nations Convention on the Law of the Sea's (UNCLOS) specific provisions relating to territorial and economic delimitation and the set of definitions of 'islands and islets'; 2) vessel flagging, ownership and piracy; 3) and joint-venture provisions for interstate common management of a natural resource.

The most common forms of conflict on the oceans in recent years may fall into three categories: territorial disputes over islands; fishing violations in territorial waters or EEZs; and piracy on the high seas. The most explosive of these issues would certainly be territorial conflict since it also overlaps into the jurisdiction of *national* defence and security involving the armed forces. Fishing violations in recent years have included greater incidents involving violence, but such disputes rarely escalate into international clashes (although historically speaking, many cases have escalated to the equivalent of cross-border fire fights). Piracy is a special case, however, since it often involves violence, creates significant economic disruption for private companies and has provoked multinational cooperation in protecting international water-ways. Often pirates are considered non-state actors, so enforcement does not typically involve jurisdictions of national defence, even if coastal navies are often involved in anti-piracy measures.

Recent disputes over the Senkaku, Takeshima and Spratly Islands in the Pacific have raised the question of distant-island ownership and the role of historical interpretation in making claims. Putting aside the notion that island claims are one way to assert hegemony over an area, the question of island territorial claims also involves specific interpretations over the legal treaties as well. Perplexingly, many of the territorial disputes between Japan and its neighbours would not fit the definition of an 'island' territorial dispute.

If a particular rock or 'islet' in the ocean is in fact deemed an 'island', then that nation may then make a claim to resources extending 200 nm from the coastal baselines, or a minimum of 125,600 square km of surrounding ocean space even if the islet is only a rock 1 metre in size. To avoid this problem, the drafters of UNCLOS were quite specific in Article 121 with respect to the definition of an island, namely that 'Rocks which cannot sustain human habitation or economic life of their own shall have no exclusive economic zone or continental shelf'.[1] Thus 'habitability' is operational in this definition and, rather surprisingly, none of the islets in the disputed territories of the Senkaku, Takeshima or the Spratlys currently sustain human habitation. Clearly, the disputes involve far more than the legal definition of the islands in question, which may explain why nations making such claims resort to historic evidence rather than contemporary legal claim. Moreover, as was mentioned in earlier chapters, these island claims are useful for stoking political fires even if cool legal resolutions may very well be available.

Another area of contention is regarding vessel-flagging and piracy on the high seas. Many vessels are able to use International Maritime Organization (IMO) regulations to their advantage to register vessels to countries with little or no regulatory enforcement so that they may conduct their activities on the ocean with impunity from enforcement or litigation. Pirates are also able to make use of such loop-holes to avoid detection by authorities, but so do vessels carrying out illegal fishing. Japan has been active in its opposition to piracy and illegal, unreported and unregulated (IUU) fisheries at United Nations forums and with bilateral partners. One such example has been the sizeable anti-piracy campaign in the Indian Ocean, conducted between 2009 and 2011, which helped secure what were becoming lawless waterways for trading vessels, especially those carry petroleum loads. Japan, Canada, the People's Republic of China (PRC), Germany, the United States and other nations participated in this operation, which made for an odd alliance of interests for a maritime cause.

The problem of reflagging of vessels and Byzantine vessel-ownership regulations make tracking of lawless ships more difficult and complicate the global measures taken for fisheries enforcement. The result is that those fishing vessels that follow laws and regulations miss out on the premium open to those who break the law, and thus make them less competitive in the open market place. Whereas anti-piracy measures have led to extensive international cooperation in policing the Malacca Straits and the Indian Ocean for trade-faring vessels, measures to combat flying flags of convenience are less noteworthy, whereby enforcement is usually voluntarily achieved by interested environmental groups rather than a centralized international body. A large-scale database of vessels registries, however, is maintained by the IMO to ensure at least a minimum of cross-border enforcement.

Lastly, one of the alternatives to the *national* claim on maritime resources is the *joint development* of such resources between two nations. Typically, this is achieved through public-private partnership agreements whereby the state and private companies mutually exploit a resource and the subsequent dividend is divided between the two. Japan and China have made such agreements with respect to oil reserves near the Senkaku Islands, but, as was seen from earlier chapters, the pedigree of such joint-venture agreements dates back to the 1950s. Even though such agreements may be touted as a new form of cooperation that can herald a new era of mutual benefit, such cooperative exploitation agreements must face the same challenges as previous ones, namely how does a government issue exploitation permits that are mutually recognized, how can private interests be balanced against public environmental costs, and how can such agreements be enforced over the long term? Oil exploitation agreements do contain an element distinct from their fisheries equivalents: the investment required to access the resource is very large. This means that territorial issues must be better defined to ensure that sizeable investments and capital do not fall into the hands of another nation-state's jurisdiction in the future. Thus, when the investments and returns for a resource peak, so too does the interest of coastal-claim enforcement.

Comprehensive security and Japanese fisheries policy in the Pacific

Several key questions were asked with regard to Japan's international fisheries policy: What were the motives behind these policies? What were the means used to achieve desired goals? Were these policies successful? How did the international environment shape Japan's capabilities in achieving these goals?

In facing the question of the *shigen mondai* (resource scarcity) and food security, Japanese policy makers have continuously opted in favour of policies that promote self-sufficiency rather than relying upon foreign imports for food needs. From early Meiji policies that promoted domestic agricultural and fisheries production to Occupation planning that fostered a system of food controls and expanded fishing rights abroad, proponents of self-sufficiency successfully laid the foundations of the food security thinking that prevails in official circles in Japan today. These ideas are entrenched to the extent that they are considered conventional wisdom amongst many bureaucrats and legislators.

These motives constitute a national interest when the strategy persists over time despite changing governments and those interests pursue general goals not necessarily restricted to a particular sectoral group. Given these two conditions, it can be said that Japan's international fisheries constituted a national interest overall, but that sectoral interests have become intertwined with national objectives in recent decades. Policies and programmes that fostered domestic food production in an effort to achieve a high level of national self-sufficiency also created with them sectoral groups which had a vested interest in the preservation of the same system that led to their creation. In a sense, national interest as it was identified at critical junctures, such as during the Meiji Restoration and the post-war Occupation, led to the development of sectoral interests which, in turn, perpetuated national policy trajectories in a relationship that may be best described as path dependent (as discussed in greater detail in Chapter 4).

This is most evident in the case of rice producers and distributors who have relied upon – and actively promoted the idea of – self-sufficiency as a desirable goal for food security. Indeed, the symbiotic relationship that rice producers and their nationwide organization, Nōkyō, have established with the Ministry of Agriculture, Forestry and Fisheries (MAFF) has ensured that self-sufficiency has remained a keystone of official planning and thinking in the bureaucracy throughout the post-war period. The same self-sufficiency strategy that has been designed for rice in the Food Agency has also been applied for fisheries policy in the closely related Fisheries Agency and, in this sense, food policy that has been adopted for rice production has guided policy for fisheries as well.

Motivated by a desire for food security and self-sufficiency in the post-war period, how did Japanese policy makers achieve these goals? The means by which these objectives were pursued have adapted to the changing world oceans regime which has moved toward a greater emphasis on norm building in multilateral fisheries institutions and away from an earlier emphasis on bilateral treaties and freedom of the seas.

Japan managed to accomplish a large measure of food security in the immediate post-war period by supporting the principle of 'freedom of the sea', which entailed few restrictions on distant-water fishing activities. With the assistance of the United States, Japan helped found the International North Pacific Fisheries Convention (1952), creating an open system in the world's most productive fishing ground, the North Pacific Ocean, while circumventing the ever-present possibility of restrictions on Japanese fishing, especially in view of Japan's position as a vanquished nation after the Pacific War. This groundbreaking treaty served as a precedent for subsequent bilateral treaties negotiated in the 1950s and 1960s with the Soviet Union, the PRC and South Korea, each in turn supporting freedom of the seas as an operating principle while opening up many fishing grounds with few restrictions on Japanese trawlers. Freedom of the seas facilitated the development of Japan's distant-water fishing industry and Japan was remarkably successful in supporting its fisheries strategy during this time.

This study examined several key points with respect to Japan's negotiating strategy for the creation of a favourable bilateral treaty system. Given the preponderance of the United States in establishing new regimes in the immediate post-war period, Japan used its special relationship with the United States to provide greater leverage when negotiating with its Asian neighbours. Japan was also effective in dealing with its Asian counterparts in a flexible, non-ideological manner, emphasizing common interests in economics, trade, fisheries and other issues that bypassed possible territorial and political impediments. Fishing agreements were often linked to other issues such as trade normalization and economic pacts. The private sector featured largely in these negotiations, especially where formal intergovernmental diplomacy was problematic or non-existent, thereby serving, in a sense, as agents of national purpose.

The support Japan secured for a system of fishing treaties based on freedom of the seas facilitated the orderly expansion of Japan's international fisheries operations and helped minimize conflict over fisheries resources for nearly three decades. Freedom of the seas enabled Japan to catch most of the fish it needed for domestic consumption and thus the Japanese government was successful in its bid to achieve a measure of guaranteed access, and thus self-sufficiency, in fisheries. Consequently, the international environment at this time can be said to have facilitated Japan's pursuit to secure a measure of food self-sufficiency.

In the mid-1970s, however, the global maritime system underwent a profound change. More nations were subscribing to a new trend in international oceans law that ran counter to the principle of freedom of the sea. Known as the ocean enclosure movement, coastal states sought to extend their control over contiguous waters, claiming jurisdiction over the exploitation of fish, minerals, oil and other resources within these waters both for economic and conservation reasons. Thus came an end to the open bilateral treaty system upon which Japan had relied for many decades.

This movement had gained enough political momentum at UNCLOS that by 1977 many states began to claim unilaterally their own EEZs of up to 200 nm from the coastline into waters that were once designated open seas. Japan was largely unsuccessful in staving off the enclosure movement's support of EEZs despite its efforts to defend the old freedom of the seas regime. The Law of the Sea negotiations in 1976–77 led to a worldwide consensus that sanctioned coastal states laying claim to fisheries resources contiguous to their national territories, essentially appropriating them from international control. Citing the need to conserve endangered fish stocks, the United States helped unilaterally initiate the EEZ system through the passage of a national fisheries law in 1976, declaring control of fishery resources within 200 nm of its coastline. This soon triggered similar claims by other states, including the Soviet Union and Canada, followed by Japan. Although the bilateral treaties continued to govern some fishing grounds in the surrounding seas, Japan lost many of its traditional distant-water fishing grounds further afield, thereby compromising Japan's national goal of food self-sufficiency.

Greater restrictions were also subsequently placed upon whaling, Bering Sea salmon fisheries, driftnet fishing on the high seas, and endangered species, often by coastal states voicing urgent resource management concerns. These enclosures and restrictions put an end to the international system of free, guaranteed access upon which Japanese distant-water fisheries had come to rely. Japan was henceforth required to negotiate new agreements in an international environment bolstered by a strong enclosure movement less amenable to distant-water fishing interests than in the early post-war period.

In light of the increasingly restrictive application of maritime law on fisheries, Japanese motives to guarantee food security were reassessed in the late 1970s. 'Comprehensive security' arose as a policy orientation that reaffirmed food self-sufficiency as a priority for government planning, and became the hallmark of Japan's foreign policy in subsequent years. The Japanese government was unwilling to abandon self-sufficiency as a goal for food security despite Japan's increasing reliance upon international imports of fish products.

Although the international oceans system had placed severe constraints upon Japan's international fisheries, Japanese officials have focused their efforts on guaranteeing the supply of fisheries resources, of which international fisheries constitute a significant share, and boosting self-sufficiency to the greatest extent possible. To achieve these ends, they have adopted a strategy of developing coastal fisheries more intensively, increasing the amount of international imports into the Japanese market, negotiating new bilateral agreements to permit access to host countries' fishing grounds, and promoting the notion of increased or free access to fisheries resources at multilateral forums and organizations.

Japan's new fishing zone was fifty times as large as its previous territorial sea and thus domestic fishermen were able to expand their fish catch easily in the short term. A significant rise in fisheries imports was also achieved, largely because the fishing capacity of these exporting nations was significantly expanded due to the declaration of their own fishing zones. In many cases, it can be said that foreign fishermen replaced the Japanese in the production of fish destined for the Japanese market.

Japan was also required to renegotiate bilateral agreements in light of the changes in domestic and international oceans law after the conclusion of the UNCLOS agreement in 1982. Renegotiated reciprocal bilateral fisheries agreements with the USSR (and later, Russia), South Korea and the PRC delineated fishing boundaries, protected Japanese domestic fishermen and established procedures for the management of fish stocks. In some cases, however, Japan had to open its own waters to foreign fishermen for the first time in its history. New bilateral agreements with the South Pacific were also established with the help of official development assistance (ODA) shaping negotiation strategies and producing outcomes favourable to Japan's distant-water fishing industry.

Finally, Japan has sought to promote increased access to fisheries resources at multilateral forums and organizations. Japan has attempted to counter efforts to restrict access to various fisheries worldwide by promoting expanded access to living maritime resources worldwide, as seen in cases such as the Food and Agriculture Organization (FAO) and the Convention for the Conservation of Southern Bluefin Tuna (CCSBT). Japan is especially cooperative with those organizations that recognize its inherent special rights as a distant-water fishing nation that has a traditional dependence upon marine resources. This has been increasingly difficult to do as support grows for the recognition of superior rights of coastal states in the management and protection of maritime resources. Nevertheless, Japan continues to defend its interests against the appropriation of maritime resources by other nations claiming coastal state precedence.

The motivation to guarantee access to scarce fisheries resources in the post-1976 international environment also helps to explain the rather perplexing stance of Japanese negotiators at International Whaling Committee (IWC) conferences today. Japanese officials wish to overturn the ban on whaling that has been in place since the 1996/97 whaling season while renewing the IWC's faith in science so that a managed fisheries regime, a 'sustainable-use' regime, can be created, instead of the current trend of irreversible moratoriums. Japanese officials have sustained the argument that whaling can be conducted on a scientific basis and have been committed to acquiring population data to support the reopening of whaling for certain species, particularly the minke whale.

As argued in Chapter 8, the whaling issue is important to Japan because it symbolizes the kind of threat that unbridled conservationism may pose to all of Japan's future fisheries and, hence, to its food security. The motivation for Japanese whaling policy is partly to protect the culture and traditions of a nation that was traditionally dependent upon whales, and still is for other marine resources. Japanese officials are also apprehensive that the whaling ban may serve as an adverse precedent that might result in similar moratoriums being applied to more valuable migratory fishing stocks, such as tuna or salmon, as they similarly become depleted due to excessive fishing pressure and inadequate management.

Domestic constituents in foreign policy making

Two key questions were asked in the Introduction of this study regarding the theoretical implications of Japan's international fisheries policy. First, did Japan employ an overarching rationale or strategy for its foreign policy in the post-war period or was it essentially ad hoc and reactive? Second, who was responsible for the development and conduct of this foreign policy?

This book has argued that Japan *did* adopt an overarching international fisheries policy throughout the post-war period – one which aimed at securing a stable supply of fisheries worldwide for the domestic market. This policy was an important component of a larger goal of comprehensive security that sought to bring a measure of resource stability by guaranteeing the supply of vital materials, such as petroleum and rice, while promoting economic interests abroad. Fisheries were deemed an indispensable food source and a necessary resource for the maintenance of Japan's food security.

Despite a colossal shift in the international fishing regime, from one that favoured distant-water fishing nations to one that gave coastal states greater control of fisheries, which compromised Japan's ability to guarantee domestically supplied fish, Japan's motive to achieve self-sufficiency in fisheries has been remarkably consistent. From the early post-war era, Japan aimed to achieve self-sufficiency in fisheries resources through the expansion of distant-water operations, and these efforts to secure a stable food supply have continued in recent times as Japan negotiates for open access to fisheries resources in multilateral agreements, consolidates reliable trading networks and fosters the highest level of self-sufficiency possible in fisheries production. Such consistency in policy strategy undermines the notion that foreign policy was strictly ad hoc in nature, since one might expect to see greater variability or randomness in the patterns of policy if this were the case.

With regard to who was responsible for developing and pursuing such a strategic policy, this study focuses on the role of national bureaucratic agencies working in concert with the distant-water fishing industry and politicians to show how national interests were achieved. As stated earlier, this study tries to investigate the institutional rationale adopted by a bureaucracy that was motivated by national strategy rather than simply focus on policy-making processes. As seen throughout this monograph, the bureaucrats deserve special attention since they are the chief architects of law and government policy relating to international fisheries, and have served primarily as the negotiators of fisheries agreements and treaties throughout the post-war period.

Moreover, the nature of resource distribution ensures that all interested parties in Japan – the bureaucrats, politicians, industry and consumers – can share the benefits of the resource without the need for competition entailed by scarcity. Bureaucrats can achieve national goals while retaining influence in the domestic economy, thereby legitimizing their ministry's normative objectives, distant-water fishing companies obtain access to new fishing grounds and the profits that this entails, politicians can retain the electoral privileges that these two groups can confer, while consumers benefit from a wider selection and stable supply of seafood products. In this particular case of Japanese foreign policy, the issue under consideration is paramount in determining the loci of political power and policy making.

Indeed, this study found little evidence that the bureaucrats have worked at cross-purposes with the interests of industry or politicians. Quite the contrary, the cases examined of fisheries negotiations with the former Soviet Union and the PRC in the 1950s, for example, demonstrate how politicians and business leaders can readily assume the role of negotiators when barriers exist to the formal participation of bureaucrats, thus augmenting the role normally adopted by the Ministry of Foreign Affairs (MOFA) and MAFF. These cases demonstrate a flexible response to a situation where the absence of normalized relations prevented formal governmental contact, yet a need existed for an international fisheries agreement. Thus, while politicians and fishing industry leaders pursued their own interests in warming relations with their erstwhile enemies and gaining access to new fishing grounds, respectively, they dually served as agents of national purpose and bureaucratic objectives to secure an important food resource. Both normative and material goals were achieved simultaneously and the interests of the bureaucrats, politicians and distant-water fishing industry were congruent in most cases studied.

One important exception, however, was witnessed during UNCLOS negotiations in the mid-1970s. As was seen in Chapter 4, MOFA and MAFF initially differed in opinion with regard to whether Japan should support the new 200-nm EEZ proposition at UNCLOS and whether Japan should claim its own EEZ jurisdiction. At first, MAFF supported the position of distant-water companies in opposing the EEZ system, but as it became clear that such a policy stance would endanger the interests of the domestic fishing industry – another competing domestic interest group – bureaucrats at both MOFA and MAFF elected to support Japan's claim to an EEZ on its eastern and northern coasts.

In this case, the interests of domestic coastal and offshore fishermen prevailed over distant-water fishing companies, largely because their industry employed more people and was responsible for significantly larger seafood production overall. It can also be argued that despite the desires of distant-water fishing interests, the new global consensus on the new EEZ system made any kind of effective opposition from Tokyo practically impossible since Japan was not willing to jeopardize its international reputation by defecting over this issue. This was a case whereby international norms prevailed over domestic interests when they came into conflict, entailing a reorientation of Japan's international fisheries strategy. Thus, with the onset of the new international maritime regime, Japan's goal of food security shifted its emphasis from a reliance strictly based on distant-water fishing companies to a greater role played by the domestic fishing industry, undermining, at the same time, the notion that the distant-water fishing industry primarily drove Japan's international fishing strategy.

The results of this dissertation also have several implications regarding the nature of Japanese foreign policy. First, international regimes and agreements largely determine the allocation and management of fishing activities, requiring nations to negotiate in bilateral or multilateral forums in order to secure the most favourable results for their respective fishing industries. Japan was able to do this in a low-profile yet effective manner, helping to influence the development of, while being profoundly affected by, the international system concerned with governing maritime resources. Second, the foreign policy strategy employed by Japanese officials followed a distinctly economistic orientation, whereby both normative and material goals were achieved simultaneously.

More than most other foreign policy issues, fisheries are profoundly affected by the international system. Since claims to maritime territories and fishing zones must be sanctified by international approval in order to be recognized, international law and custom essentially define the rights and rules within which the fishing industry can conduct their operations. This means that policy must be devised by national negotiators – MOFA and MAFF in the case of Japan – and thus the system in which resources are managed and allocated is a product of negotiating strategies and transferable international legal norms. As seen earlier, Japanese negotiators rarely adopted a leadership role in such negotiations, preferring instead to make incremental gains in negotiations on a case-by-case basis while being motivated by an overarching strategy of food security. This was a case where Japanese foreign policy was proactive, low profile and mostly effective in achieving its stated goals.

At the same time, Japan frequently relied upon the leadership role of its close ally the United States in helping to shape international norms and regimes in its favour, most notably in the long-term defence of the freedom of the seas regime worldwide. When Washington switched its support to national enclosures and resource protection in the case of salmon, pollack and whales, however, Japanese officials defended their distant-water fishing rights – although with mixed success. Thus, Japan relied upon the favours of the United States when their interests coincided, but was willing to defy Washington when those same interests became threatened.

It can also be argued that Japanese foreign policy on international fisheries relations was distinctly economistic. Such relations, however, were not neo-mercantilist insofar as the distant-water fishing industry was not solely responsible for the development of fisheries policy – although it did derive significant material gains from it.[2] Instead, the bureaucracy pursued fisheries policy in line with the economic strategy evident in many other industries, while consistent with the Yoshida Doctrine which sought to minimize foreign expenditures so as to devote resources to the development of domestic strategic industries and, later, the pursuit of comprehensive security. Economic goals, such as food security, were deemed on par with traditional military and politico-strategic goals of other nations. Thus the bureaucrats pursued a strategy that entailed both normative nation-building goals while deriving material benefits for the fishing industry and Japanese consumers.

Ocean diplomacy: future developments and Japanese responses

As we have seen, changing legal regimes had the overall impact of placing restrictions on Japanese distant-water fishing, with a negative impact on its self-sufficiency and food security. In years to come, the most likely factor that will make it difficult for Japan to maintain, let alone increase, its food self-sufficiency is environmental change – a change that has been leading to greater scarcity of fishery resources worldwide. This has been compounded by growing demand for fish products in recent years, which is partly the result of changing consumer tastes in the developed world in favour of fish, as well as rapidly growing populations in the developing world. International cooperation in fisheries management will become increasingly difficult to encourage as fishing resources dwindle and their allocations to each fishing nation decline with growing resource scarcity.

Environmental change poses the greatest threat to any long-term strategy Japan, or any country for that matter, may use to guarantee a stable supply of fish with a high level of self-sufficiency. Many new fisheries biology and population studies have suggested that the world fish catch has been declining in recent years as a result of overfishing, and this has had a negative impact on ocean habitat and marine ecosystems. Naturally, fewer fish in the sea will mean fewer landings for fishermen in the near future.

Overfishing is not the only threat to sustainable fisheries. Global climate change will have an impact on ocean currents and could have drastic, unforeseeable consequences on marine ecosystems and food chains. The Japanese government's National Research Institute of Fisheries Engineering predicts that some of Japan's domestic fish catch might decline by as much as 70 percent over the next one hundred years due to global warming.[3] If so, domestic fisheries will suffer a blow that might result in a precipitous, irreversible decline in self-sufficiency and the loss of livelihoods for many people in coastal communities.

The problem of scarcity is compounded with growing demand for fish. The global population is growing by 77 million people a year and the United Nations estimates that it will increase from 6.1 billion to 9.3 billion people by the year 2050.[4] Such demographic leaps will necessarily entail greater competition and demand for food sources, including fisheries. Moreover, developed countries now also have a growing appetite for fish as it becomes more fashionable to eat, and highly recommended by nutritionists for its low-fat and high Omega-3 fatty-acid content.[5] Greater pressure will be brought to bear on the world's fisheries as an alternative source of protein at a point when even now the FAO currently estimates that 75 percent of the world's commercial fish stocks are fully exploited, overexploited or significantly depleted.[6] Greater demand for fish will create greater relative scarcity, with an impact on prices and distribution, whereby richer countries will have to pay more to secure fisheries resources that are reaching their biological limits.[7]

Fisheries are growing into an issue with broader public interest than any time before and this will likely entail greater support for environmental non-governmental organizations' (NGOs) efforts to protect and conserve world fisheries that have, until now, been allowed to operate outside of the public's view. As was noted in earlier chapters, environmental groups have been shifting the focus of their campaigns to target deep sea fishing and the threats posed by overfishing. The Johannesburg Earth Summit's widely supported proposal to set rules for sustainable world fisheries is but one example of the rising importance of fisheries issues to environmental groups.

Despite the growing scientific and public concern for overfishing and declining fish stocks, Japanese bureaucrats continue to be highly sceptical about predictions of fish stock collapse or claims of overfishing resulting in environmental damage. There seems to be a widespread belief in the inexhaustibility of the oceans which still pervades official thinking in Japanese circles, and many bureaucrats and industry representatives are reluctant to talk about declining fish populations. On the contrary, Japanese fishing industry publications repeatedly stress that a greater fishing effort, not smaller, is required to boost fishing production since there are more fish in the ocean than scientists estimate.[8] Moreover, if a fish stock were to collapse in one part of the world, they can almost always be found in another ocean somewhere. As a fishing nation, Japan's interests are shored up by the insurance offered by the vastness of the oceans and Japan's ability to find new fishing grounds or new importers for supplies that dwindle elsewhere. Thus, even if fish stocks are actually in decline worldwide and overfishing threatens their long-term sustainability, Japan's ability to supply the fish it needs is not yet perceived to be under immediate threat.

Instead, Japanese officials stress that non-restrictive, rule-based international regimes, based upon scientific evidence and recommendations, must be applied to manage fisheries properly. Japan is keen to continue working with international organizations for the management of fisheries, especially if they recognize the special rights inherent to distant-water fishing nations. As can be seen with Japan's policy with respect to the IWC, it seems unlikely that Japan will attempt to defect from international organizations even when their interests are perceived to be slighted or contradicted. In a sense, Japan feels it has more to gain from a commitment to international order over fisheries than by 'going it alone' through unilateral action.

How will Japan respond to greater scarcity in fisheries and greater scrutiny by environmentalists? It will likely continue to counter efforts by environmentalists and concerned coastal states to have marine species listed for special protection through moratoriums, trade restrictions or marine reserves.

Japanese officials will also be unlikely to support experimental fishing programmes that give property rights to coastal fishermen, as has been tried in
New Zealand and Iceland, since such systems appropriate control from
common management, create economic incentives that encourage restrictions
on fish production and often result in a fish catches with higher cost to
consumers.[9]

Instead, Japan will likely continue to promote the concept of 'sustainable'
fisheries, which derive their sustainability from the fact that scientific evidence
will be used to bolster management decisions. However, critics add that
without proper management systems that limit incentives to catch more fish
than one's competitors (i.e. the tragedy of the commons), no amount of scientific evidence, no matter how good its quality, will lead to sustainable fisheries. The real problem with creating sustainable fisheries today is fisheries
management systems that rely upon common use and control of fisheries, and
not the quality of scientific evidence and recommendations feeding into these
management models.

Placing emphasis on scientific evidence will also shift decision making away
from the influence of world public opinion, which Japanese bureaucrats and
politicians have been largely unsuccessful in shaping, and will give greater
power to a select group of specialists and fisheries experts in deciding the
allocation of increasingly scarce fisheries resources. Such a shift in emphasis
may make decision making less transparent to non-specialists, while increasing the technicality of resource-allocation disputes in the future. Japan may
also use the preponderant position and size of its market for fish products to
help shape future decision-making processes of international fisheries organizations, since non-compliance by the leading fish-consuming nation in the
world will undermine any alternative fisheries management strategies.

The goal of food security has been a consistent feature of Japanese international fisheries policy throughout the post-war period and will continue to
play an important role in the future as well. The means by which this was
achieved have adapted to changing world circumstances and legal regimes,
which have mostly been successful in guaranteeing a reliable supply of fishery
resources to the Japanese market, even if this has entailed some declines in
the rate of self-sufficiency. The greatest challenge that Japanese self-sufficiency
in fisheries will face in the near future will be the risks inherent in environmental change and declining fish populations. To be successful at achieving
some measure of security with a stable supply of resources, Japanese officials
will once again have to prove themselves capable of adapting to new rules and
systems of fishery management.

Notes

1 United Nations, *United Nations Convention of the Law of the Sea*, 10 December
1982, Article 121, United Nations Treaty Series, 66.
2 Neo-mercantilism, as it is used here in this dissertation, is understood in the
context of policy making as when private industries retain enough influence over

government essentially to determine the course of national policy. It can be argued, however, that such policies were 'mercantilist' if the term is more broadly understood as referring to a strong state that pursues policies that bolster the wealth, and thus power, of a nation (especially through maximizing exports and minimizing imports). For example, The *Oxford Companion to Politics of the World* describes it thus: 'Mercantilist politics varied from country to country, but their principal objectives were similar: to maximize the power of the sovereign and the well-being of his or her subjects. Under mercantilism power and wealth were synonyms not opposites' (*Oxford Companion to Politics of the World*, second edn, Joel Krieger *et al.*, ed. New York: Oxford University Press, 2001, 535.

3 *Kyodo News*, 'Fish Catches in Japan to Decline by up to 70% Due to Global Warming', 28 August 2005. The institute made the prediction based on the assumption that water temperatures will have risen by 1.4–2.9°C by 2100. It studied thirty-four varieties of fish and possible changes in catches at fishing ports. Ports in Nagasaki and Kagoshima prefectures facing the East China Sea are expected to face 30–70 percent declines in catches of Japanese jack mackerel, chub mackerel, red sea bream and a few other varieties.

4 United Nations Environment Programme (UNEP), *Global Environment Outlook 3*. London: Earthscan, 2002, 33.

5 Charles Clover, *The End of the Line: How Over-Fishing is Changing the World and What We Eat*. London: Ebury Press, 2004, 2.

6 Even this is considered a conservative estimate. UNEP, *Global Environment Outlook 3*, 184.

7 It can also be argued, however, that the present situation of population decline in Japan will make it easier to achieve food self-sufficiency in the future than would be otherwise. Nonetheless, any such benefits that may be accrued from this are likely to be outweighed by greater competition among world fishing fleets, including Japanese, for increasingly scarce fisheries resources that would be entailed by an increasing world population.

8 See, for example, the 2003 series in the JFA's publication, *Isaribi*, regarding the under-reporting of tuna populations worldwide. Although most fisheries scientists outside of Japan are warning of excessive fishing pressure on, and possible population collapse of, many tuna stocks, this series of articles argued that tuna stocks worldwide are not only healthy but capable of sustaining an even higher catch than at present.

9 Although the Individual Transferable Quota (ITQ) system has been widely hailed by fisheries economists as a practicable system that minimizes the problems associated with commons management, Japanese fisheries economists have remained highly sceptical of its applicability in Japan.

Bibliography

Clover, Charles. *The End of the Line: How Over-Fishing is Changing the World and What We Eat*. London: Ebury Press, 2004.

United Nations. *United Nations Convention of the Law of the Sea*. United Nations Treaty Series, 1982.

UNEP (United Nations Environment Programme). *Global Environment Outlook 3*. London: Earthscan, 2002.

Appendix 1: The international fisheries of Japan

Profiles of industry and policy makers

This appendix will examine Japan's international fishing industry and introduce the key policy makers who helped shape the political and legal environment in which it operated. It begins with an outline of fisheries classifications within Japan, defining the distant-water fisheries relative to the other domestic fishing sectors. This will illustrate the importance of international fishing in relation to other industries and in terms of Japan's overall food supply. Japan's international fisheries will then be profiled for 1975, a year after historic peak production for distant-water fleets and one year before the institution of the 200-nm exclusive economic zones (EEZs), and some comparisons will be made with the state of this industry for the year 2000. Several key trends in the pelagic fisheries over the last four decades will also be examined, including the identification of key fishing grounds in foreign waters and the most valuable fisheries within the distant-water fishing industry. This will help provide more extensive geographic background to the political and economic fisheries relations that were discussed in the book.

Whereas distant-water fisheries provided a significant share of production in 1975, this supply was steadily displaced by an increasing reliance upon international imports by 2000. The increasingly open nature of the Japanese market as well as rationalization within the distant-water fishing industry help to explain some of the domestic factors that led to this shift in production. The role that changing international regimes had in delineating new fishing boundaries and the impact this had upon domestic production is an important theme of the book.

The latter part of the appendix will provide a short introduction to the key companies operating in the industry along with an overview of the Japan Fisheries Association, the official industry lobbyist which also serves as a broad-based interest group. Finally, key bureaucratic and political agencies will be profiled, starting with the Ministry of Foreign Affairs and the Ministry of Agriculture, Forestry and Fisheries. The role of politicians and prime ministers will also be evaluated.

Japanese fisheries classifications

Japan's long tradition of coastal, beach and inland-water fishing stretches back many centuries and it would be difficult to overstate the importance of both fish and the fishing industry to the Japanese. Three distinct types of fisheries emerged in the early twentieth century, distinguished both by their areas of operation and by their techniques: coastal, offshore and distant water.

Coastal fisheries are the oldest form of fishing in Japan. Such fisheries are largely conducted by village associations or diverse small-scale individual enterprises characterized by techniques requiring low-capitalization such as simple equipment and small-scale production. Their methods of operation include beach seines, lift nets, set nets or traps, gill nets, and hook-and-line fishing within the customary 12-nm territorial zone, as well as shellfish and seaweed collection, and inland fresh-water catches.

Offshore fishing ranges from just outside the territorial sea to hundreds of miles from the coast. Due to the higher cost of equipment and mechanized boats capable of operating at such ranges, offshore fishing is largely conducted by companies and special cooperative associations that can afford such an investment. These fisheries engage in large-scale purse-seine fishing, two-boat power trawling, and line-and-pole fishing techniques.

As a consequence of the high cost and technical sophistication of refrigeration systems, larger diesel engines and ships, distant-water fisheries did not develop extensively until the late 1920s. Even then, only large companies such as Taiyo Gyogyō and Nippon Suisan (Nissui) could afford the capital investment and organizational expertise needed to be profitable. Such fisheries have primarily operated motorized factory ships and trawlers off the coasts of Russia, China's Kwantung Peninsula, South Korea, Taiwan and in the South Pacific, as well as mothership operations in the Sea of Okhotsk, the Bering Sea and the North Pacific high seas.

International fisheries are defined here as the combination of distant-water fishing and imports of fish products from abroad. Since Japanese distant-water fleets often operate near the coasts of other countries and imports also originate from fisheries operating in foreign waters, international fisheries provide a sum of fishing effort on the high seas and in fishing grounds near foreign countries.

The Fisheries Agency defines Japanese fisheries according to areas of operation, types of fishing techniques and gear used, and species targeted, which results in a potentially confusing array of statistical classifications. In general, coastal fisheries are conducted within a few kilometres of the Japanese coastline, whereas offshore fisheries can range out as far as a few hundred kilometres. Distant-water operations are conducted on the high seas or in the shallow waters near the coasts of other countries.

Another pronounced division among the major fisheries sectors is the type of fishing rights employed. Distant-water fishing rights are allocated by licences issued by the Ministry of Agriculture, Forestry and Fisheries (MAFF). In most cases the licence period is set at five years and MAFF retains the right to continue or cancel the licence according to its management scheme.[1] When catch quotas are decided annually through negotiations under bilateral agreements, as in the case with salmon fisheries with the former USSR, the licence period is reduced to one year, but renewable annually.[2] Coastal and offshore fisheries, on the other hand, are determined not by licences but through traditional rights based upon allocations that in some cases have existed since the Tokugawa era. Fisheries cooperatives headquartered in respective areas of jurisdiction oversee these fisheries and are responsible for the practical implementation of a system aimed at mitigating excessive competition for limited fisheries resources. Rights to fishing grounds frequently coincide with fishing village boundaries, although there have been cases where a fisheries cooperative manages fishing grounds littoral to another fishing community.[3] The width of the zone under management varies from place to place, but is on average about 1 km from the shore.[4]

International fisheries profile in 1975

In terms of the overall importance of the fisheries, Figures A.1 and A.2 represent the percentage of distant-water fisheries catch relative to the other fisheries in both value and volume, respectively. As Figure A.1 shows, the distant-water catch in 1975 was roughly equal in value to the other two main marine fisheries, offshore and coastal, each capturing close to one-third of the marine catch. The value of distant-water, offshore and coastal fisheries was ¥480 billion, ¥530 billion and ¥510 billion, respectively, or the equivalent of US$6.4 billion, $7.0 billion and $6.9 billion in 2002 real terms. The distant-water fishing industry was close in overall value to other such industries as iron imports, the construction value of all factories, warehouses and power stations combined, or one-tenth the value of all petroleum imports.[5] Moreover, the imports of fish products into Japan totalled 710,000 metric tonnes, valued at ¥385 billion, or about 80 percent of the value of the distant-water fisheries.[6] Since such fisheries originated near the coasts of other states, it can be seen that a majority of roughly 45 percent of the fish consumed in Japan originated in international waters. In terms of volume, the distant-water fisheries caught 3.2 million metric tonnes of fish, totalling 31 percent of the weight for the year's catch. This catch constituted several species of more valuable fish and shellfish on a per kilogram basis than coastal or offshore fisheries, such as tuna, salmon and shrimp. Fisheries conducted offshore had the largest catch, representing over 40 percent of overall marine landings with a total haul of close to 4.5 million metric tonnes.

The MAFF definition of distant-water fisheries has changed over time.[7] While the overall effect of such changes on statistical catch rates is negligible, it is important to keep this in mind when considering the range of operations undertaken by this class of fisheries. According to the MAFF *Statistical*

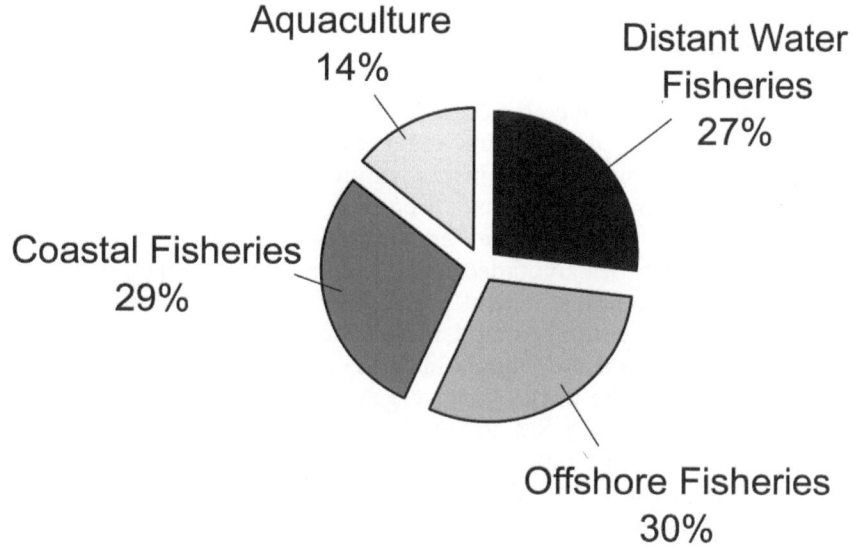

Figure A.1 Percentage of value for Japanese marine fisheries, 1975
Source: MAFF. *Gyogyō yōshokugyō seisan tōkei nenpō, 50 nendo* (Statistical Yearbook of
Fisheries and Aquaculture Production, 1975). Tokyo: Nōrinsuisanshō tōkei jōhōbu, 1976

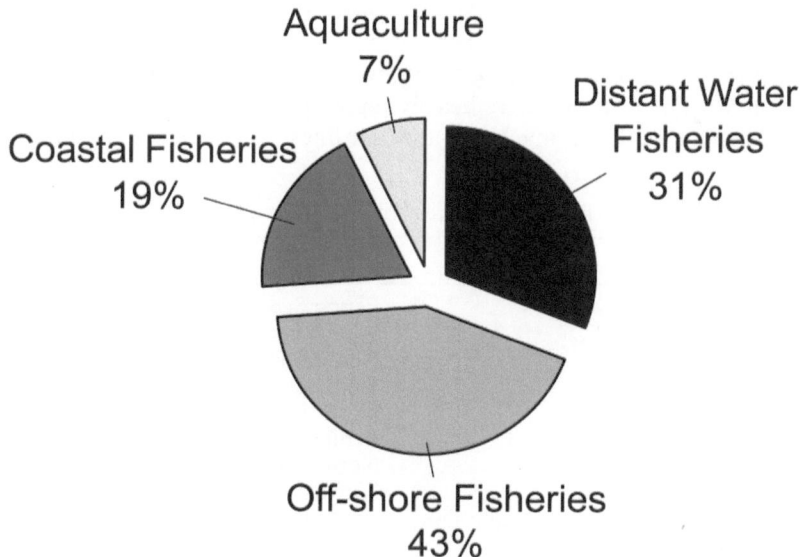

Figure A.2 Percentage of volume for Japanese marine fisheries, 1975
Source: MAFF. *Gyogyō yōshokugyō seisan tōkei nenpō, 50 nendo* (Statistical Yearbook of
Fisheries and Aquaculture Production, 1975). Tokyo: Nōrinsuisanshō tōkei jōhōbu, 1976

Yearbook of Fisheries and Aquaculture Production, the distant-water fisheries were defined as illustrated in Table A.1.

A general, if somewhat incomplete, illustration of the principal distant-water fishing grounds may be drawn for 1975. Table A.2 shows that a remarkable 40 percent of the volume of all marine fisheries was caught within 200 nm of other countries. One major feature of the distant-water fisheries was that most of its fishing grounds were not located on the high seas but close to the coasts of other countries. Distant-water fishing effort near the coasts of other countries amounted to 3.7 million metric tonnes, as opposed to 5.5 million metric tonnes caught near Japan.[8] Of this amount, 38 percent was caught in waters littoral to the United States, while 37 percent originated from USSR coastal waters.[9] Thus, while Japan's distant-water fisheries operated in all oceans of the world, they were mostly concentrated in the North-West Pacific (i.e. the Bering Sea and the Sea of Okhotsk), where more than 30 percent of the total maritime catch originated in 1975.[10] The chief species targeted in these fishing grounds was the Alaskan pollack, representing three-quarters of all catches – a trend that continued well into the 1980s.[11]

Japan's most productive fishing grounds were found in the North-West Pacific, where nearly 9 million tonnes of marine fish were caught, or 83 percent of their overall catch in 1973.[12] While these figures do not show the extent of distant-water fishing (since this classification includes Japanese domestic waters), it does suggest the importance of the North-West Pacific as a fishing-ground ecosystem to Japanese fishers. The rest of the Pacific as a whole accounted for another 12 percent.[13]

The principal species of fish targeted by Japanese ships depended upon the fishing areas. Three species were heavily represented in overall marine catch in 1975: Alaskan pollack, mackerel and sardines. These three species alone

Table A.1 Distant-water fisheries and areas of operation in 1975

Fisheries	Areas of operation
Mothership dragnet	Bering Sea
North Sea trawlers	Waters north of Japan
South Sea trawlers	New Zealand, Atlantic and Africa
Western trawlers	South China Sea
Mothership salmon trawlers	North Pacific
Mothership crab trawlers	Bering Sea, northern waters
North Pacific snow crab	North Pacific
North Sea gillnets	Waters north of Japan
Distant-water skipjack tuna pole and line	South Pacific
Mothership tuna longline	South Pacific
Distant-water tuna longline	South Pacific

Source: MAFF. *Gyogyō yōshokugyō seisan tōkei nenpō, 51 nendo* (Statistical Yearbook of Fisheries and Aquaculture Production, 1975). Tokyo: Nōrinsuisanshō tōkei jōhōbu, 1976; and Yohoji Asada, Yutaka Hirasawa and Fukuzo Nagasaki. *Fishery Management in Japan*. FAO Fisheries Technical Paper 238. Rome: FAO, 1992

Table A.2 Distant-water fishing grounds and volumes of fish catch

Fishing grounds (within 200nm)	Volume ('000mt)
United States	1,410
Soviet Union	1,396
People's Republic of China	152
South and North Korea	241
Australia	12
New Zealand	80
Canada	21
Others	432
Total	3,744
Within Japanese waters	5,503
Percentage within 200 nm of other states	40%

Source: MAFF. *Suisan hakusho, Shōwa 51 nendo* (Fisheries White Paper, 1976). Tokyo: Nōrin tōkei kyōkai, 1977, 6

accounted for more than 40 percent of all marine fisheries, including coastal and offshore.[14] For distant-water fisheries, however, Alaskan pollack, salmon and tuna were the most valuable. According to Table A.3, mothership drag-net, North Sea and 'Hoku-ten' distant-water trawlers targeted Alaskan pollack (83 percent of catch) while Western trawl operations west of 130° East longitude focused on mackerel (21 percent) and hake (18 percent). Mother-ship salmon trawlers and salmon driftnetters fished for various salmon and saltwater trout, while crab trawlers focused on snow crabs.[15] Skipjack tuna pole-and-line operations and tuna longliners fished mostly for skipjack tuna, long fin tuna, sailfish and yellow tail.

An examination of the value of fish catches reveals some interesting trends. Table A.4 shows that the most lucrative fishery in 1975 was clearly the tuna longline operations that mostly plied the South Pacific. The value of this catch accrued not so much from the volume of the catch but from the price

Table A.3 Distant-water fisheries target fish (figures are % of catch volume)

Mothership-type trawlers	
Number of people working	5,155
Catch	Alaskan pollack (83%)
	Flat fish (13%)
Distant-water trawlers (northern, 'Hoku-ten', southern and shrimp trawlers)	
Northern catch	Alaskan pollack (76%)
Hoku-ten catch	Alaskan pollack (86%)
Southern and shrimp catch	Various (29%)
	Horse mackerel (23%)
	Hake (18%)
	Squid (9%)

Western trawlers and draggers

Trawlers catch	Various (60%)
	Yellow croaker (20%)
Draggers catch	Various (28%)
	Squid (13%)
	Flatfish (12%)
	Shiroguchi (11%)

Mothership salmon trawlers

Number of people working	6,644
Catch	Salmon (52%)
	Saltwater trout (48%)

Mothership crab trawlers

Number of people working	508
Catch	Snow crab (100%)

Skipjack tuna pole and line (distant-water, offshore and coastal)

% of catch by vessels 50 tonnes or over (largest category)	88%
Catch	Skipjack tuna (76%)
	Long fin tuna (16%)

Mothership tuna longline

Catch	Sailfish (38%)
	Yellow tail (32%)

Distant-water tuna longline (Atlantic)

Catch	None

Distant-water tuna longline (Pacific and Indian Oceans)

Catch	None

Salmon driftnet fishing[1]

Catch	Salmon (60%)
	Saltwater trout (40%)

North Pacific snow crab[2]

Notes: [1] MAFF statistics sometimes also refer to this as the Northern Ocean gillnets fishery; [2] Figure not included in MAFF 1975 statistical table.
Source: MAFF. *Gyogyō yōshokugyō seisan tōkei nenpō, 51 nendo* (Statistical Yearbook of Fisheries and Aquaculture Production, 1975). Tokyo: Nōrinsuisanshō tōkei jōhōbu, 1976

fetched by tuna in the domestic market, which was the second highest at an average of ¥664 per kilogram. The next most valuable fishery, the distant-water trawlers, were not so much a high-priced fishery as they were efficient, with a sizable catch amounting to 1.7 million metric tonnes of fish. Many of the most valuable fisheries operations in Table A.4 had significant distant-water components, such as squid and salmon fisheries. Moreover, Table A.5 reveals that the two most valuable fisheries by far on a per kilogram basis were the salmon and tuna fisheries, their value nearly doubling that of the third-ranked price for squid. Both of these fisheries were sizable distant-water operations.

Studies of Japanese fisheries frequently tend to emphasize catch volume as a measure of the importance of the components of the industry. As can be seen in Figure A.3, an examination of fisheries catches by volume in 1975 tends to over-emphasize the role of purse-seiners that largely operated within Japanese waters, and underestimates the overall importance of more valuable fisheries such as tuna longline that only represented 3 percent of total fisheries catch.

Purse-seiners largely targeted sardines which were an abundant fish, but not very valuable in the domestic market, since it was largely used as agricultural feed and fertilizer and not for human consumption. A composite graph is constructed in Figure A.4 which plots fisheries in terms of volume (bars) and value (lines). This shows that the most lucrative fisheries, i.e. high value for catch, are again the tuna, squid and salmon fisheries.

In summary, Japan's distant-water fishing operations, which focused on the North-West Pacific, especially in fishing grounds near the Soviet Union and the United States, provided an important supply of fish products to the domestic market. The most valuable of these fisheries were tuna, squid,

Table A.4 Rank in order of overall value for major fisheries in 1975

		(x 10,000 tonnes)	*(x ¥100 million)*	*(¥/kg)*
1	Tuna longline	29	1,939	664
2	Distant-water trawlers	168	1,478	88
	North Sea trawler	53	402	
	Hoku-ten trawler	86	478	
	South Sea trawler	29	599	
3	Squid jigging	40	1,395	346
4	Purse-seiners	204	1,373	67
	Large and medium-sized (mainly sardine)	157	no price	
5	Salmon fisheries	15	1,016	677
	Mothership	3	179	
	Gill net	6	368	
	Set net	6	453	
6	Skipjack tuna pole and line	32	900	279
7	Offshore trawlers	137	788	58
8	Shellfish and seaweed	40	634	157
9	Western trawlers	21	561	266
10	Whales (number of)	13,427	364	
11	Mothership trawlers	80	336	42
12	Saury dip net	22	247	114

Note: Not separated according to area of operation, i.e. coastal, offshore or distant-water.
Source: MAFF. *Gyogyō yōshokugyō seisan tōkei nenpō, 51 nendo* (Statistical Yearbook of Fisheries and Aquaculture Production, 1975). Tokyo: Nōrinsuisanshō tōkei jōhōbu, 1976

Table A.5 Rank in order of fish value (¥/kg)

	Rank	(¥/kg)	(10,000 tonnes)	(¥100 million)
1 (5)	Salmon fisheries	677	15.0	1,016
2 (1)	Tuna longline	664	29.2	1,939
3 (3)	Squid jigging	346	40.3	1,395
4 (6)	Skipjack tuna pole and line	279	32.3	900
5 (9)	Western trawlers	266	21.1	561
6 (8)	Shellfish and seaweed	157	40.3	634
7 (12)	Saury dip net	114	21.6	247
8 (2)	Distant-water trawlers	88	167.8	1,478
9 (4)	Purse-seiners	67	203.9	1,373
10 (7)	Offshore trawlers	58	137	788
11 (11)	Mothership trawlers	42	79.6	336

Note: Numbers in parentheses represent ranking by total value.
Source: MAFF. *Gyogyō yōshokugyō seisan tōkei nenpō, 51 nendo* (Statistical Yearbook of Fisheries and Aquaculture Production, 1975). Tokyo: Nōrinsuisanshō tōkei jōhōbu, 1976

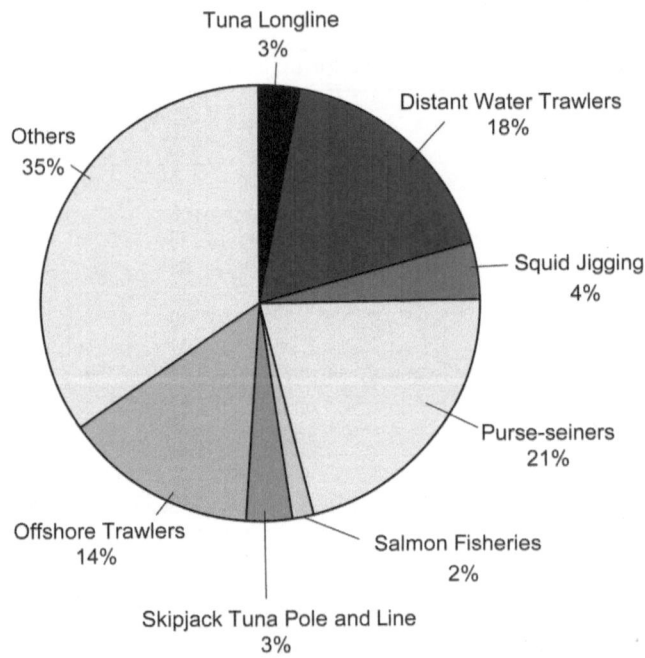

Figure A.3 Major marine fisheries by catch volume 1975
Source: MAFF. *Gyogyō yōshokugyō seisan tōkei nenpō, 51 nendo* (Statistical Yearbook of Fisheries and Aquaculture Production, 1975). Tokyo: Nōrinsuisanshō tōkei jōhōbu, 1976

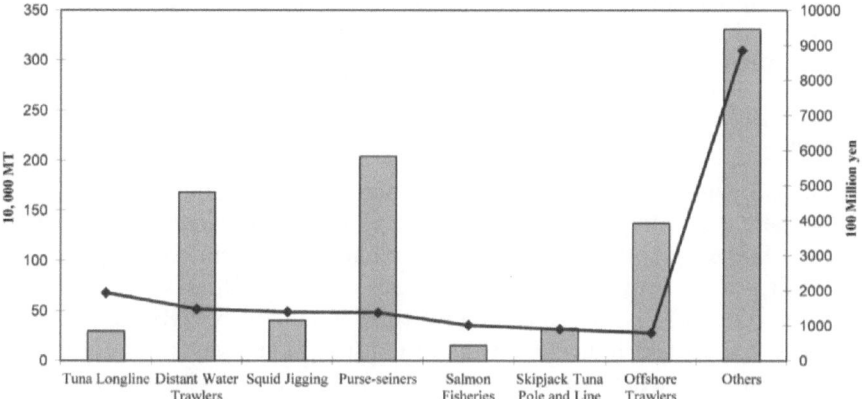

Figure A.4 Major fisheries in 1975
Note: Bars represent volume, lines represent value.
Source: MAFF. *Gyogyō yōshokugyō seisan tōkei nenpō, 51 nendo* (Statistical Yearbook of Fisheries and Aquaculture Production, 1975). Tokyo: Nōrinsuisanshō tōkei jōhōbu, 1976

salmon and Alaskan Pollack. In terms of volume, Alaskan pollack, sardines and mackerel were heavily represented.

Other trends for Japanese fisheries in 1975

Imports

There was a ten-fold increase in fish imports to Japan in the 1965–75 period. Japan imported a total of 710,000 metric tonnes of fish, valued at a total of ¥385 billion in 1975, amounting to 38 percent of fisheries supply in terms of value.[16] Thus, imports were a very important component in Japan's food balance sheet even as early as 1975, before the introduction of the 200-nm EEZs.

These imports were higher-quality fishery products destined for consumption, including frozen shrimp, squid and octopus, with a swift rise in the total amount of tuna.[17] Some 35 percent of the frozen shrimp originated from India, Indonesia and China, while the lion's share of tuna imports came from South Korea. Of the volume of products imported, 29 percent were fresh/frozen products, 26 percent processed fish paste and 20 percent canned or salted fish.[18]

Exports

Japan was also an active supplier of fish products to the world market, especially to North America. The export industry amounted to a ¥133 billion business, or 7 percent of the value of all marine fisheries and about one-third of the value of fisheries imports. Half of all exported goods were canned seafood, mostly canned mackerel and tuna. Much of the canned mackerel was destined for the Philippines, while most of the canned tuna was exported to the United States, valued at ¥36 billion and ¥28 billion, respectively.[19]

Demand

Concurrent with the rise in household incomes throughout the 1960s and 1970s, there was a trend of increasing fish consumption per capita, generally aimed at higher-quality products. Half of all animal protein consumption was fish, with a daily average intake of 75g per person.[20] Domestic expenditure upon fish products rose 85 percent between 1960 and 1975.[21] Rising prices for fresh and frozen fish indicate a shift in consumption away from traditional products such as canned and salted fish, the price for which was steadily declining in the 1970s.[22] There was also a 150 percent increase in demand for minced fish paste products (*surimi*), made from the abundant Alaskan pollack and often consumed in the traditional Japanese dish *oden* or in hot pots (*nabe*). Moreover, the rise in popularity of sashimi (fresh raw fish), tempura and fried fish led to increasing demand for fish products and suggested a significant change in tastes. In many ways these changing appetites demanded greater diversification and general improvement in the quality of fish products.[23]

Costs

The high seas fisheries faced rising costs in the mid-1970s as a result of a radical rise in oil prices and an increase in labour wages. The cost of oil increased three-fold from 1970 prices, while the price of ropes and netting doubled.[24] This was a major expenditure for distant-water fishing companies, whose fleets had for years relied on cheap fuel costs to make profitable runs to fishing grounds located great distances from Japan.

Another significant cost to distant-water operations was labour. A review of expenditures for the distant-water large trawlers showed that by the mid-1980s wages accounted for nearly one-third of all costs, the largest item by far.[25] Rising wages were witnessed throughout the 1970s as well. This was largely attributable to the rise in opportunity costs for fisheries labour, as rising education and alternative job opportunities in the major cities created a significant drain on the labour pool in fishing communities.[26] Fishermen made on average 81 percent of a farmer's wage, and 86 percent of that of a general labourer.

Lastly, the general ageing of workers in the fishing industry compounded the labour problem. This required young workers to replace older ones as the latter retired or moved to other forms of employment. However, many young people did not find such work as appealing as the opportunities afforded by an increasingly industrialized society offering lucrative wages in manufacturing and services. Higher wages had to be offered to entice them, which meant that replacement labour costs rose steadily. Moreover, the overall number of people employed in 1975 was roughly equal to that of a decade earlier.[27] This suggests that while the replacement rate generally balanced the retirement rate, there was also an ageing trend within the labour force as the many young workers recruited in the 1950s started their third decade of employment. These factors created more expensive conditions for the high seas fisheries in 1975 relative to earlier decades, when cheap labour and fuel costs enabled the expansion of fisheries and the provision of a reliable fisheries supply from abroad.

Japan's international fisheries profile, 2000

By 2000, Japan's overall maritime food production dropped to a total of nearly 6 million metric tonnes, which placed it as the world's third-largest producer. As can be seen in Figure A.5, the People's Republic of China (PRC) overtook Japan as the world's largest producer in 1988 and has held that position ever since.[28] By contrast, Japan is still the world's largest market for fish products, as will be discussed in greater detail later. Despite the rising importance of beef, pork and chicken in the Japanese diet, 40 percent of all animal protein consumed remains fish products.[29] This is a significant drop, however, when compared to figures a few decades before. In the 1955–60

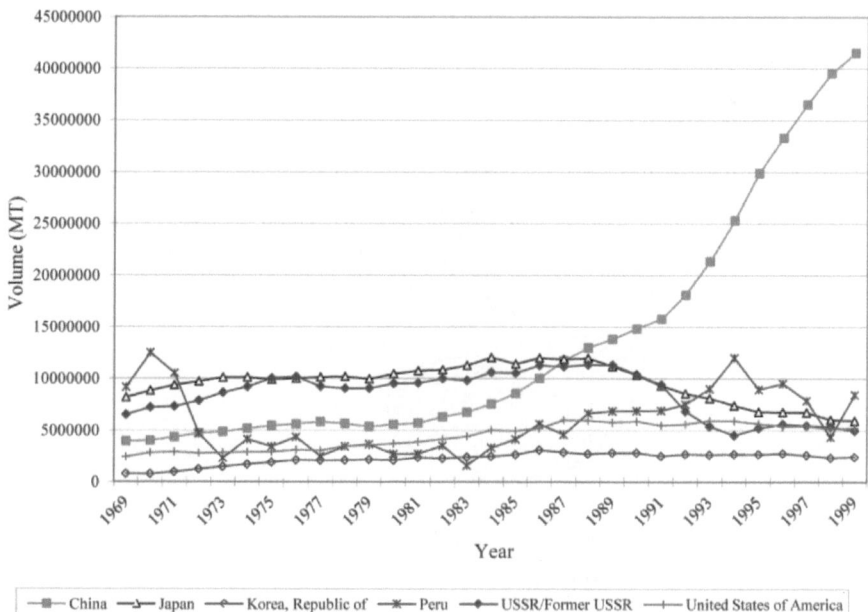

Figure A.5 Fisheries production by major states, 1969–99
Source: FAO Fishstat database, www.fao.org/fi/statist/FISOFT/FISHPLUS.asp

period, Japanese dependence on fish products as a source of animal protein reached a high of 78 percent.[30]

The sectoral composition of Japan's marine fish catch can be seen in Figure A.6. Whereas the distant-water fisheries once occupied a full third of overall fish catch in years of peak production, the overall percentage of the distant-water fisheries in terms of value and volume dropped to 9 percent and 11 percent, respectively, whereas domestic production jumped to nearly half of overall value and 38 percent of volume. This general trend of the declining importance of distant-water fishing in Japan's overall maritime production can also be seen in Figure A.7. Whereas the steady increase in overall production in the early post-war era until the mid-1970s can be attributed to rising distant-water production, the trend after the mid-1970s has been progressive decline of production for this sector. Offshore fisheries experienced a significant rise in production volume in the late 1970s to 1990 due to sizable catches of Japanese pilchard and sardines, but these figures have declined in recent years due to the cyclical abundance of such species. It should also be noted that, as previously, Japanese pilchard and sardines are mostly used as fertilizer or feed and thus do not have much impact on Japan's food stock budget.

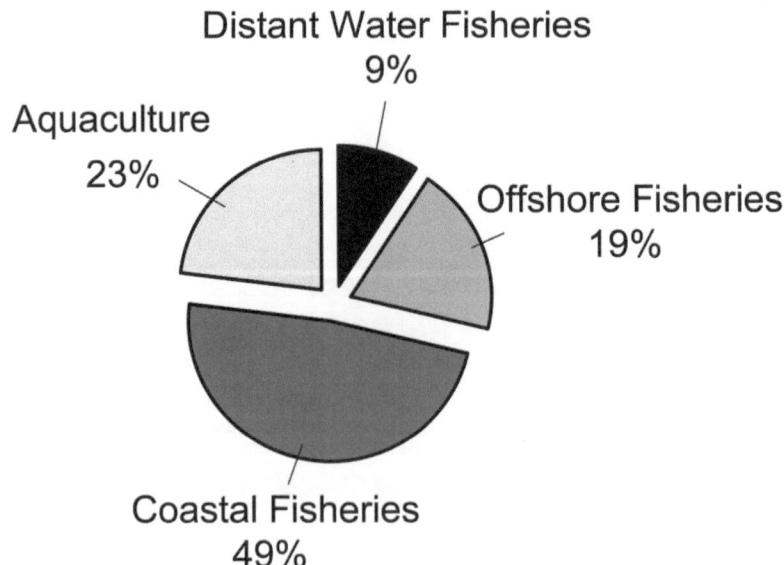

Figure A.6 Percentage value of Japanese marine fisheries, 2000
Source: MAFF. *Suisan hakusho, Heisei 13 nendo* (Fisheries White Paper, 2001). Tokyo: Nōrin tōkei kyōkai, 2002

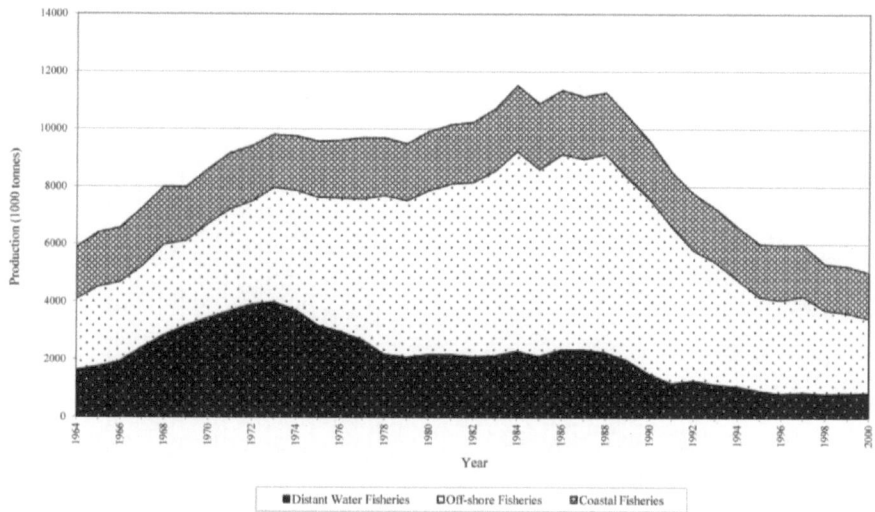

Figure A.7 Marine fisheries production by sector, 1952–2000 (volume)
Source: MAFF. *Gyogyō yōshokugyō seisan tōkei nenpō* (Statistical Yearbook of Fisheries and Aquaculture Production). Tokyo: Nōrinsuisanshō tōkei jōhōbu, 1957–2000

Fisheries imports

Japan is the largest importer of seafood products, accounting for 26 percent of world import value in 2001.[31] Fisheries imports to Japan have been steadily increasing in recent years in both volume and value, with a pronounced jump in both areas in the late 1980s.

Although Figure A.8 does not offer values in real terms, it nevertheless reliably indicates a strong correlation between rising value and volume over three decades. The total demand for fisheries products has fluctuated between 8 and 9 million tonnes in recent years.[32] The precipitous rise in value after 1985 seems likely to be the result of the Plaza Accord that same year, which revalued the yen against the US dollar and led to a consequent rise in import values in terms of the US dollar. Whereas the Plaza Accord may explain a sudden rise in import values, it does not, however, sufficiently explain the subsequent steepness of rising imports in later years. Thus, rising demand seems to be the only likely explanation for rising import values after 1985.[33]

About half of Japan's edible fish supply comes from foreign countries on a raw material basis. This is due to shifting demand from medium- to high-price fish that cannot be met by domestic production, as well as a general decline in domestic production.[34] The quantity of imports was 3.5 million tonnes, while their value was $1.7 billion.[35] This represents a five-fold increase in volume from 1975 figures. Figure A.9 shows the value of fish imports relative to other major import products, such as crude oil and forestry products.

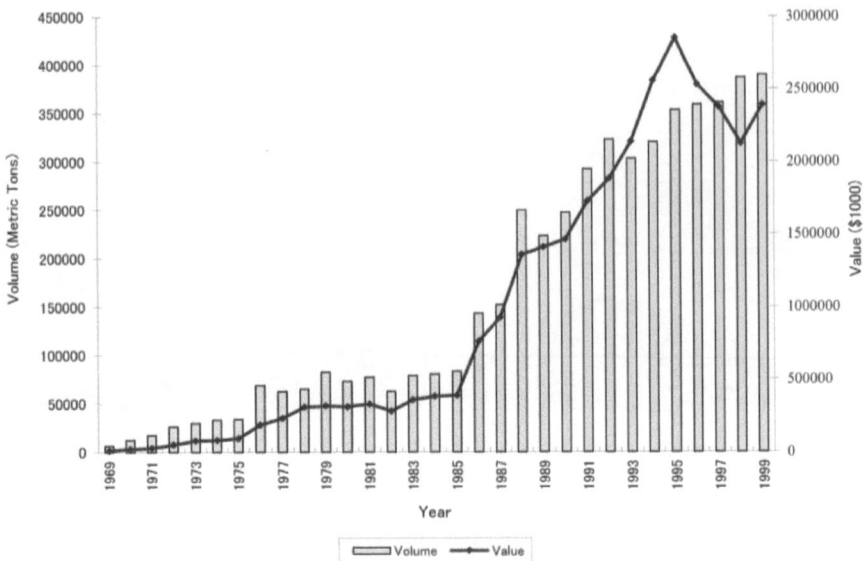

Figure A.8 Japan fisheries imports by volume and value, 1969–99
Source: FAO Fishstat database, www.fao.org/fi/statist/FISOFT/FISHPLUS.asp

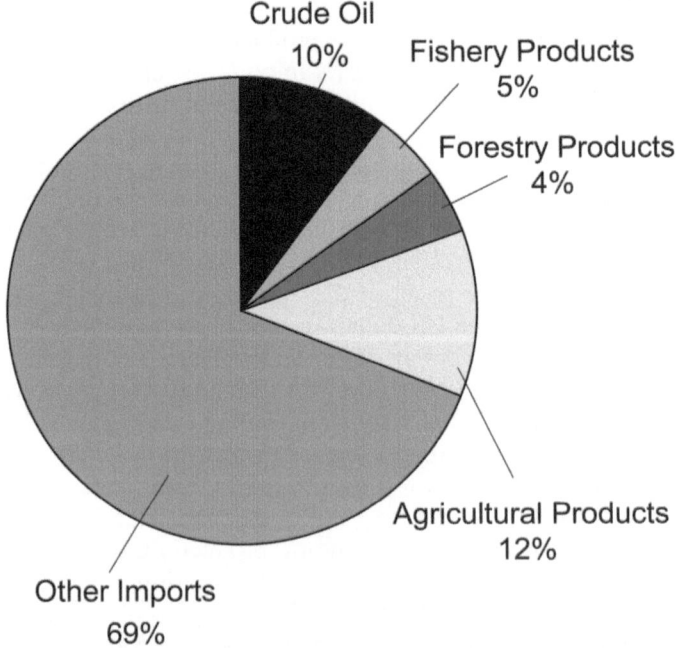

Figure A.9 Imported products as a percentage of total import value
Source: MAFF Online Statistical Abstracts, www.maff.go.jp/abst/abindex.html

In recent years China has been Japan's largest trading partner in terms of import value, closely followed by the United States and Russia. In 2000, Japan imported close to 15 percent of its marine fisheries in value from China, and another 10 percent from the United States, 8 percent from Russia and another 7 percent each from South Korea, Thailand and Indonesia.[36] In terms of fish products as represented in Figure A.10, shrimp captures 19 percent of this total, whereas skipjack/tuna represented 13 percent, salmon 7 percent and crab 6 percent.[37]

Despite the rise in import value and volume in recent years, the price of fish products has been remarkably stable over the last four decades. Figure A.11 shows the relative value of fish catch in real terms and indicates that whereas distant-water fisheries have undergone significant price fluctuation, the overall price of fisheries goods has stayed close to ¥250/kg, with the exception of the 1977–84 period. This period coincides with the restructuring of distant-water fisheries after the implementation of EEZs, which was discussed earlier in the book. Price stability for fisheries products is one of the policy aims of MAFF, and these figures suggest that this goal has mostly been achieved over the years, despite possible price shocks due to changing international circumstances.

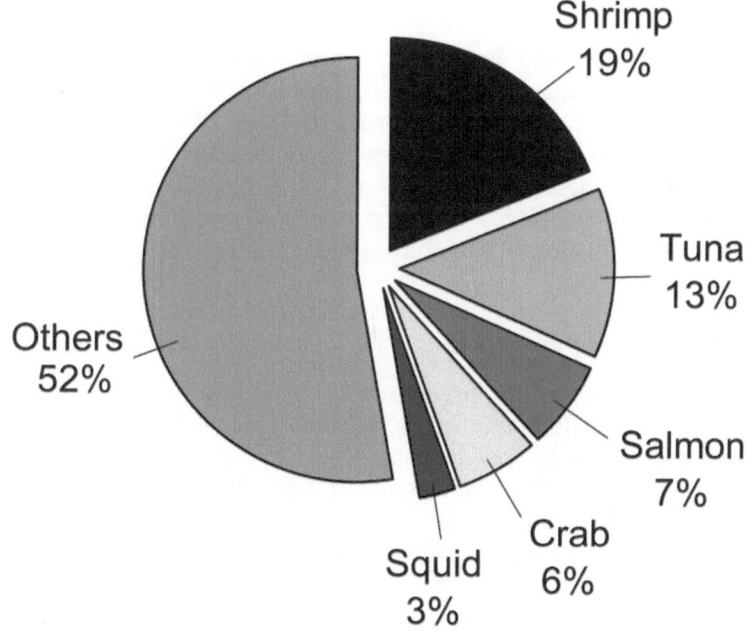

Figure A.10 Value of imports by product 2000
Source: MAFF. *Suisan hakusho, Heisei 13 nendo* (Fisheries White Paper, 2001). Tokyo: Nōrin tōkei kyōkai, 2002

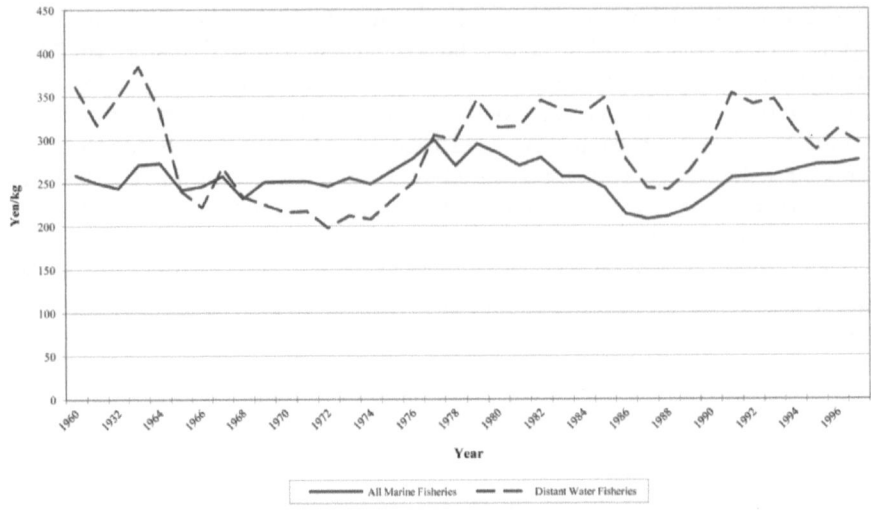

Figure A.11 Marine product real prices, 1960–98 (1990 base year)
Source: MAFF. *Gyogyō yōshokugyō seisan tōkei nenpō* (Statistical Yearbook of Fisheries and Aquaculture Production). Tokyo: Nōrinsuisanshō tōkei jōhōbu, 1957–2000

The rise of imports into Japan in the past two decades has had a detrimental impact upon Japan's self-sufficiency ratios for fish. Whereas Japan was self-sufficient for 79 percent of fish products from 1960–88, self-sufficiency declined to 53 percent in 2000, which is less than many other key commodities such as rice, vegetables and chicken.[38] The government of Japan set a target for 65% self-sufficiency by the year 2012, which was to require an increase in the production of seafood up to 5.26 million tonnes.[39]

Distant-water fishing companies

Whereas coastal fisheries are largely run by local fisheries cooperatives, several large companies dominate Japan's distant-water fishing industry. Of the five major companies in this industry, the two largest are Taiyō Gyogyō (renamed Maruha in 1993) and Nissui. In 1990, Taiyō Gyogyō had annual sales of $4.4 billion while Nissui's sales amounted to $3.2 billion.[40] These companies employed close to 3,500 people each in 1990, a figure that has declined in recent years.[41] Their operations range from harvesting to processing and distribution, suggesting a highly integrated corporate structure within the fishing industry.

Until the mid-1980s, trawling operations were the largest division within both Taiyō Gyogyō and Nissui. These companies were capable of the capital investment and organizational expertise necessary to turn international expeditions into profitable ventures. Distant-water companies established a widespread network of onshore bases that offered services in processing and freezing as well as port facilities for repairs and refuelling near all major fishing grounds. Such facilities were frequently offered in joint venture arrangements as part of a general effort, as described by Kiyoshi Kojima and Terutomo Ozawa, in securing access to primary resources.[42]

In recent years, however, their import divisions have become the focus of these companies' operations, essentially transforming their roles from fishing companies to food importers. Through a variety of means Taiyō and Nissui have been able to offer a stable supply of high-quality fisheries products to Japan despite the shrinkage of harvesting operations. Such means have included chartering Japanese vessels and crew to coastal-state fishing companies; ensuring that they are the largest purchaser of fisheries goods from traders, thus retaining the position of core buyer; and establishing joint ventures with host countries, typically in non-equity – 'over the side' – contractual relationships.[43] Moreover, the strong Japanese appetite for high-quality seafood and the increased buying power of the Japanese yen after the Plaza Accords contributed significantly to creating economic conditions supportive of trade in fish products.[44]

Japan Fisheries Association

When making appeals to the government concerning fisheries policy, these companies have usually availed themselves of the Japan Fisheries Association (JFA): a broad-based, non-governmental advocacy organization with close links to the Fisheries Agency, with a membership of more than 400 associations, private companies and individuals.[45] Since the Association of Fisheries Cooperatives (*Zengyoren*) serves as the chief spokesperson for coastal and offshore fisheries, the JFA has traditionally worked closely with distant-water fisheries companies in the articulation of their international interests.

Since inshore and offshore fishermen's interests have not always coincided with those of the distant-water fisheries companies, the JFA has served as a forum for policy debate and consensus forging among various fishing interest groups before approaching the government with policy preferences. The strength of this organization has changed over time, coinciding with the importance of distant-water fisheries to the domestic market. Nonetheless, the JFA remains an important spokesperson for distant-water fishing companies.[46]

One of the JFA's chief aims is to establish committees and discussion panels regarding government policies in an effort to coordinate views in the industry and to lobby Diet members and the bureaucracy.[47] The JFA has twenty-five standing committees on a range of fisheries concerns such the environment, distribution and processing, Japan-PRC relations, foreign fishing policy and Japan-South Korea relations.[48] The honorary president of the JFA until 2004 was former Prime Minister Suzuki Zenkō, an influential politician with strong links to the fishing industry and someone who will be discussed further below.

Government of Japan

This section will briefly examine the structure and function of the bureaucratic agencies responsible for international fisheries policy. Decision making is largely the result of coordination between two principal agencies: the Ministry of Foreign Affairs (MOFA) and MAFF (including its subsidiary, the Fisheries Agency). The Fisheries Agency, in particular, offers specialist advice on technical fisheries problems, which gives it added influence in the development of Japan's international fisheries policy.

Ministry of Foreign Affairs

The ministry primarily responsible for the conduct of Japanese foreign policy is MOFA. The section within the ministry that deals with international fisheries policy is the Fisheries Division, Economic Affairs Bureau. It is relatively small compared to other divisions within more prominent areas such as the Bureaux for North America or Economic Cooperation, but it nonetheless conducts a wide range of activities including treaty negotiations, representing Japan at international organizations, hosting international fisheries conferences, and maintaining the everyday affairs of bilateral fisheries relations such as repatriating seized vessels and settling catch quotas.[49]

While MOFA is not always the chief architect of foreign policy, it serves as the mouthpiece by which domestically derived policies are articulated within Japan's larger foreign policy concerns. As seen in numerous studies by such authors as Tsuneo Akaha, Gerald Curtis and Alan Rix, MOFA works in cooperation with other branches of government that hold domestic jurisdiction of the issue area in question in order to coordinate Japan's international relations.[50] For example, the funding, allocation and monitoring of overseas development assistance are coordinated with the Ministries of Finance, and Industry, Trade and Investment, among others, whereas MOFA works with the Ministry of Education in the promotion of Japanese language and culture abroad.[51]

In the case of international fisheries, MOFA has a close relationship with MAFF, or more specifically the Fisheries Agency within this ministry, in coordinating foreign policy. While the relationship between the two ministries has occasionally been contentious, from United Nations Conference on the Law of the Sea (UNCLOS) negotiations in the 1970s to Japan's current stance on whaling, they have largely been successful in keeping their differences out of view and in presenting a unified case at international forums.

Ministry for Agriculture, Forestry and Fisheries

International fisheries policy is largely determined by the Fisheries Agency, which is under the supervision of MAFF. The ministry provides personnel and budgetary funding to the Fisheries Agency and closely monitors the agency's activities.[52] MAFF also has several other divisions that relate to Japan's international fisheries strategy, including the General Food Policy Bureau, the International Affairs Department (primarily responsible for official development assistance (ODA) and affairs relating to the World Trade Organization), and the Statistics Department. The stated mission of MAFF is to achieve the sustainable development of the agricultural, forestry and fisheries industries through securing a stable supply of safe foods, developing rural areas and making use of the three industries' multifunctionality.[53]

The Fisheries Agency is responsible for both domestic and international fisheries under the Basic Fisheries Law of 2001. The stated aims of the Fisheries Law are two-fold: 1) securing a stable supply of fishery products; and 2) the sound development of the fishing industry to promote the appropriate conservation and management of marine resources.[54] International fisheries policy, including decisions relating to whaling, ODA and bilateral treaty negotiations, falls under the mandate of the International Affairs Division and the Far Seas Division. The primary objective of the Fisheries Agency is to provide a stable supply of marine products by sustainably securing natural marine resources, promoting fish farming, protecting fishing grounds and cooperating in international resource management.[55]

Although MOFA assumes the responsibility of representing Japan at most large international meetings and conferences, it is not unusual for their representatives to be accompanied by officials from the Fisheries Agency. In some cases, Fisheries Agency officials serve as the official representative of Japan, particularly if the international meeting is small or special expertise is required. It is also customary for a Japanese official from the Fisheries Agency to be appointed as the head of the Fisheries Division of the Food and Agriculture Organization (FAO).[56]

Politicians

Normally, politicians from both governing and opposition parties have not become actively involved in international fisheries policy making but have instead deferred such responsibilities to the bureaucracy. However, when an international fisheries issue gains status, as in the case of the UNCLOS negotiations in the 1970s or normalization talks with the Soviet Union and South Korea (1952–65), politicians have been known to take a stance and the prime minister sometimes exercises his executive authority to drive policy in a desired direction. The most influential catalysts in the political realm are policy groups (*zoku*) within the governing Liberal Democratic Party (LDP) and personal initiatives by the prime minister.

Fisheries zoku

The fisheries policy group within the LDP is a small one relative to other, better-known groups, such as the agricultural or transportation *zoku*. It has traditionally been concerned with domestic fisheries and issues affecting voters living in rural fishing communities, which is not surprising considering that the risk of becoming entangled in international fisheries questions does not translate into many constituency votes.[57] Several subgroups have emerged in recent years, including a pro-whaling league that actively promoted the whaling cause at the International Whaling Commission annual meeting in Shimonoseki, although their support for whaling is typically more symbolic than actually influential.[58]

Prime ministers

Prime ministers do not normally involve themselves in the day-to-day affairs of fisheries policy unless an issue arises that is of immediate national importance. However, there are some notable cases in the past when prime ministers became directly involved in international fisheries policy. One such case was the leading role played by Hatoyama Ichirō in the normalization of relations and associated conclusion of a fisheries treaty with the Soviet Union in 1956, as was discussed in Chapter 3. Another case of interest is the promotion of the LDP fisheries *zoku*'s leader, Suzuki Zenkō, to prime minister in 1980. Prime Minister Suzuki was nicknamed *Sakana-san* ('Mr Fish') before assuming leadership of the country, and was known as a strong supporter of the fishing sector throughout his career. He originated from Iwate Prefecture, a base for the distant-water fishing industry, and served as a fisheries technocrat in the JFA before starting his political career in the Diet.[59]

Conclusion

Japan's international fisheries have constituted an important sector of fisheries production throughout the post-war period. The North-West Pacific contains many important fishing grounds for Japanese fishermen, especially in the waters around Japan and within 200nm of Russia (the former Soviet Union) and the United States. Many commercially valuable species were supplied by the distant-water fishing industry.

Whereas international fisheries production was largely based upon distant-water catches until the mid-1970s, this largely shifted to a reliance on imports by 2000. The Japanese economy is still very reliant upon fishing grounds near the coasts of other countries, but the production of fisheries products has shifted from Japanese distant-water fishing companies to overseas producers who now export to Japan. Consequently, large fishing companies that formerly oversaw distant-water fleets active in oceans around the world have now shifted the emphasis of their business to importing fisheries products from a vast network of foreign producers.

There are several domestic bureaucratic and political organizations that have helped shape the political and legal environment in which their industry counterparts have had to cope with these changes. Although MOFA is the ministry primarily responsible for the conduct of Japanese foreign policy in general and international fisheries policy in particular, the Fisheries Agency of MAFF serves in an advisory capacity, assists in formulating fisheries policy concerned with foreign countries and distant waters, and is the primary destination of lobbying efforts by the fishing industry, including the JFA, which chiefly articulates the interests of the distant-water fishing companies. MOFA and the Fisheries Agency have largely worked in concert, but there are instances when consensus regarding foreign policy preferences was difficult to achieve between these organizations – at times to the detriment of Japan's overall negotiating stance at international forums. On occasion, the prime minister and fisheries policy *zoku* can help break such deadlocks and offer leadership for the resources of government to pursue a preferred policy on international fisheries.

The international oceans system underwent several changes in the post-war period, which entailed significant consequences for Japan's international fisheries. The shift in Japan's international fisheries from distant-water production to a heavy reliance on imports also coincides with the international systems shift from an open fisheries regime to a closed one in later years. While the international system is not the only factor that created such a shift, it is an overridingly important one in explaining the determination of Japan's policy/ means in the search for food security.

Notes

1 Yohoji Asada, Yutaka Hirasawa and Fukuzo Nagasaki. *Fishery Management in Japan*. Food and Agricultural Organization Fisheries (FAO) Technical Paper 238. Rome: FAO, 1992, 12.
2 Ibid., 12.
3 Such cases are rare, however.
4 Asada *et al.*, *Fishery Management in Japan*, 9.
5 Office of the Prime Minister. *Japan Statistical Yearbook 1977*. Tokyo: Bureau of Statistics, 1978.
6 Ibid.
7 These changes have been made for statistical purposes at least.
8 MAFF. *Suisan hakusho, Shōwa 51 nendo* (Fisheries White Paper, 1976). Tokyo: Nōrin tōkei kyōkai, 1977, 4.
9 Ibid., 4.
10 Ibid., 3.
11 Ibid.
12 The Northwest Pacific is defined in accordance with the Food and Agriculture Organization's classification of ocean areas.
13 FAO. *Yearbook of Fisheries Statistics: Catches and Landings, Vol. 36*. Rome: FAO, 1973.
14 MAFF, *Suisan hakusho, Shōwa 51 nendo*.
15 One point of interest here is that 1975 MAFF statistics declare no other fish catches, despite a later study by Linda Paul of Earthwatch (www.earthtrust.org/dnpaper/contents.html) which found a very high level of by-catch for these fisheries.
16 MAFF, *Suisan hakusho, Shōwa 51 nendo*, 9.

17 Ibid., 9.
18 Ibid., 9.
19 MAFF, *Suisan hakusho, Shōwa 51 nendo*, 22.
20 Ibid., 4.
21 Ibid.
22 Asada *et al.*, *Fishery Management in Japan*, 11.
23 MAFF, *Suisan hakusho, Shōwa 51 nendo*, 5.
24 Ibid., 13.
25 Olav Stokke. 'Transnational Fishing: Japan's Changing Strategy', *Marine Policy*, vol. 15 (July 1991): 234.
26 Masatoshi Yorimitsu. 'On the Labour Market of Off-shore Fisheries', *Hitotsubashi Journal of Social Studies*, vol. 13 (November 1981): 23.
27 MAFF, *Suisan hakusho, Shōwa 51 nendo*, 16.
28 Recently, however, serious questions have been raised about the accuracy of Chinese fisheries statistics as well as those of other major fishing nations. See, for example, Reg Watson and Daniel Pauly. 'Systematic Distortions in World Fisheries Catch Trends', *Nature*, no. 414 (29 November 2001): 534–36; and *The Economist*, 'Fishy Figures' (29 November 2001). Despite the questionable nature of such statistics, the spectacular rise in Chinese fishing effort is undisputed, as is their position as the largest fishing nation today.
29 MAFF. *Suisan hakusho zusetsu, Heisei 13 nendo* (Fisheries White Paper, Illustrated Version, 2001). Tokyo: MAFF Nōrin tōkei kyokai, 2002, 95.
30 MAFF. *Annual Report on Japan's Fisheries, Fiscal 1991*. Tokyo: Ministry of Agriculture, Forestry and Fisheries, 1991, 19.
31 Organisation for Economic Co-operation and Development (OECD). *Review of Fisheries in OECD Countries: Policies and Summary Statistics, 2002*. Paris: OECD, 2002, 37.
32 OECD. *Review of Fisheries in OECD Countries: Policies and Summary Statistics, 2003*. Paris: OECD, 2003, 301.
33 I would like to thank Professor Kazutoshi Kase at the Institute of Social Sciences, the University of Tokyo, for raising this point.
34 OECD, *Review of Fisheries in OECD Countries, 2003*, 301.
35 Ibid., 302.
36 MAFF, *Suisan hakusho zusetsu, 12-nendo*, 90.
37 Ibid., 89.
38 MAFF, *Suisan hakusho zusetsu*, 97; and John T. Sproul. 'Effects of North Pacific 200-Mile Exclusive Economic Zone Marine Management Policy on Japanese Seafood Production, Trade and Food Security', *Bulletin for the Faculty of Fisheries, Hokkaido University*, vol. 43, no. 3 (1992): 145.
39 OECD, *Review of Fisheries in OECD Countries, 2003*, 44, footnote 31.
40 *Japan Company Handbook*. Tokyo: Tōyō Keizai Shinpōsha, 1990.
41 Ibid.
42 Kiyoshi Kojima and Terutomo Ozawa. *Japan's General Trading Companies: Merchants of Economic Power*. Paris: OECD Development Centre, 1984.
43 Olav Schram Stokke. 'Transnational Fishing: Japan's Changing Strategy', *Marine Policy*, vol. 15 (July 1991): 239–41.
44 Sproul, 'Effects of North Pacific 200-Mile Exclusive Economic Zones', 142.
45 Japan Fisheries Association, publicity material.
46 Although its power has diminished in recent years with the decline of distant water fisheries.
47 Japan Fisheries Association, www.suisankai.or.jp/index_e.html.
48 Ibid.
49 Interview with Director, International Fisheries Division, Economics Bureau, Ministry of Foreign Affairs (26 February 2004).

50 For example, see Tsuneo Akaha. *Japan in Global Ocean Politics.* Honolulu: University of Hawaii Press, 1985; Gerald Curtis, ed., *Japan's Foreign Policy after the Cold War: Coping with Change.* Armonk, NY: M.E. Sharpe, 1993; and Alan Rix, *Japan's Foreign Aid Challenge: Policy Reform and Aid Leadership.* London: Routledge, 1993.

51 For authoritative studies on Japanese aid policy formulation and coordination, see Alan Rix, *Japan's Foreign Aid Challenge*; and Robert Orr. *The Emergence of Japan's Foreign Aid Power.* New York: Columbia University Press, 1990.

52 Keiko Hirata. 'Beached Whales: Examining Japan's Rejection of an International Norm', *Social Science Japan Journal*, vol. 7, no. 2 (2004): 190.

53 MAFF. 'Guide to MAFF'. www.maff.go.jp/e_guide/004.htm.

54 OECD, *Review of Fisheries in OECD Countries, 2002*, 18.

55 MAFF. 'Guide to MAFF'. www.maff.go.jp/e_guide/028.htm.

56 Interview with Purwito Martosubroto, Fisheries Resource Officer, Fisheries Division, FAO (28 January 2002).

57 Discussion with Liberal Democratic Party official, International Bureau (26 July 2003).

58 Hirata, 'Beached Whales', 192.

59 Sandra Tarte. *Japan's Aid Diplomacy and the Pacific Islands.* Melbourne: Asia Pacific Press, National Centre for Development Studies, 1998, 66.

Index